These safety symbols are used in laboratory and field investigations in this bo[ok]
ing of each symbol and refer to this page often. *Remember to wash your hand[s]*

MW00700435

PROTECTIVE EQUIPMENT Do not begin any lab without the prope[r]

 GOGGLES Proper eye protection must be worn when performing or observing science activities that involve items or conditions as listed below.

 APRON Wear an approved apron when using substances that could stain, wet, or destroy cloth.

 SOAP Wash hands with soap and water before removing goggles and after all lab activities.

GLOVES Wear gloves when working with biological materials, chemicals, animals, or materials that can stain or irritate hands.

LABORATORY HAZARDS

Symbols	Potential Hazards	Precaution	Response
DISPOSAL	contamination of classroom or environment due to improper disposal of materials such as chemicals and live specimens	• DO NOT dispose of hazardous materials in the sink or trash can. • Dispose of wastes as directed by your teacher.	• If hazardous materials are disposed of improperly, notify your teacher immediately.
EXTREME TEMPERATURE	skin burns due to extremely hot or cold materials such as hot glass, liquids, or metals; liquid nitrogen; dry ice	• Use proper protective equipment, such as hot mitts and/or tongs, when handling objects with extreme temperatures.	• If injury occurs, notify your teacher immediately.
SHARP OBJECTS	punctures or cuts from sharp objects such as razor blades, pins, scalpels, and broken glass	• Handle glassware carefully to avoid breakage. • Walk with sharp objects pointed downward, away from you and others.	• If broken glass or injury occurs, notify your teacher immediately.
ELECTRICAL	electric shock or skin burn due to improper grounding, short circuits, liquid spills, or exposed wires	• Check condition of wires and apparatus for fraying or uninsulated wires, and broken or cracked equipment. • Use only GFCI-protected outlets	• DO NOT attempt to fix electrical problems. Notify your teacher immediately.
CHEMICAL	skin irritation or burns, breathing difficulty, and/or poisoning due to touching, swallowing, or inhalation of chemicals such as acids, bases, bleach, metal compounds, iodine, poinsettias, pollen, ammonia, acetone, nail polish remover, heated chemicals, mothballs, and any other chemicals labeled or known to be dangerous	• Wear proper protective equipment such as goggles, apron, and gloves when using chemicals. • Ensure proper room ventilation or use a fume hood when using materials that produce fumes. • NEVER smell fumes directly. • NEVER taste or eat any material in the laboratory.	• If contact occurs, immediately flush affected area with water and notify your teacher. • If a spill occurs, leave the area immediately and notify your teacher.
FLAMMABLE	unexpected fire due to liquids or gases that ignite easily such as rubbing alcohol	• Avoid open flames, sparks, or heat when flammable liquids are present.	• If a fire occurs, leave the area immediately and notify your teacher.
OPEN FLAME	burns or fire due to open flame from matches, Bunsen burners, or burning materials	• Tie back loose hair and clothing. • Keep flame away from all materials. • Follow teacher instructions when lighting and extinguishing flames. • Use proper protection, such as hot mitts or tongs, when handling hot objects.	• If a fire occurs, leave the area immediately and notify your teacher.
ANIMAL SAFETY	injury to or from laboratory animals	• Wear proper protective equipment such as gloves, apron, and goggles when working with animals. • Wash hands after handling animals.	• If injury occurs, notify your teacher immediately.
BIOLOGICAL	infection or adverse reaction due to contact with organisms such as bacteria, fungi, and biological materials such as blood, animal or plant materials	• Wear proper protective equipment such as gloves, goggles, and apron when working with biological materials. • Avoid skin contact with an organism or any part of the organism. • Wash hands after handling organisms.	• If contact occurs, wash the affected area and notify your teacher immediately.
FUME	breathing difficulties from inhalation of fumes from substances such as ammonia, acetone, nail polish remover, heated chemicals, and mothballs	• Wear goggles, apron, and gloves. • Ensure proper room ventilation or use a fume hood when using substances that produce fumes. • NEVER smell fumes directly.	• If a spill occurs, leave area and notify your teacher immediately.
IRRITANT	irritation of skin, mucous membranes, or respiratory tract due to materials such as acids, bases, bleach, pollen, mothballs, steel wool, and potassium permanganate	• Wear goggles, apron, and gloves. • Wear a dust mask to protect against fine particles.	• If skin contact occurs, immediately flush the affected area with water and notify your teacher.
RADIOACTIVE	excessive exposure from alpha, beta, and gamma particles	• Remove gloves and wash hands with soap and water before removing remainder of protective equipment.	• If cracks or holes are found in the container, notify your teacher immediately.

Your online portal to everything you need

connectED.mcgraw-hill.com

Look for these icons to access exciting digital resources

 Video

Audio

Review

Inquiry

WebQuest

Assessment

Concepts in Motion

Mc Graw Hill **Education**

INTERACTIONS OF LIFE

i SCIENCE

Glencoe

INTERACTIONS OF LIFE

SCIENCE

Glencoe

Snow Leopard, *Uncia uncia*

The snow leopard lives in central Asia at altitudes of 3,000 m–5,500 m. Its thick fur and broad, furry feet are two of its adaptations that make it well suited to a snowy environment. Snow leopards cannot roar but can hiss, growl, and make other sounds.

The McGraw·Hill Companies

 Education

Send all inquiries to:
McGraw-Hill Education
8787 Orion Place
Columbus, OH 43240-4027

ISBN: 978-0-07-888018-6
MHID: 0-07-888018-1

Printed in the United States of America.

4 5 6 7 8 9 10 11 DOW 15 14 13

Authors and Contributors

Authors

American Museum of Natural History
New York, NY

Michelle Anderson, MS
Lecturer
The Ohio State University
Columbus, OH

Juli Berwald, PhD
Science Writer
Austin, TX

John F. Bolzan, PhD
Science Writer
Columbus, OH

Rachel Clark, MS
Science Writer
Moscow, ID

Patricia Craig, MS
Science Writer
Bozeman, MT

Randall Frost, PhD
Science Writer
Pleasanton, CA

Lisa S. Gardiner, PhD
Science Writer
Denver, CO

Jennifer Gonya, PhD
The Ohio State University
Columbus, OH

Mary Ann Grobbel, MD
Science Writer
Grand Rapids, MI

Whitney Crispen Hagins, MA, MAT
Biology Teacher
Lexington High School
Lexington, MA

Carole Holmberg, BS
Planetarium Director
Calusa Nature Center and
Planetarium, Inc.
Fort Myers, FL

Tina C. Hopper
Science Writer
Rockwall, TX

Jonathan D. W. Kahl, PhD
Professor of Atmospheric Science
University of Wisconsin-
Milwaukee
Milwaukee, WI

Nanette Kalis
Science Writer
Athens, OH

S. Page Keeley, MEd
Maine Mathematics and
Science Alliance
Augusta, ME

Cindy Klevickis, PhD
Professor of Integrated Science
and Technology
James Madison University
Harrisonburg, VA

Kimberly Fekany Lee, PhD
Science Writer
La Grange, IL

Michael Manga, PhD
Professor
University of California, Berkeley
Berkeley, CA

Devi Ried Mathieu
Science Writer
Sebastopol, CA

Elizabeth A. Nagy-Shadman, PhD
Geology Professor
Pasadena City College
Pasadena, CA

William D. Rogers, DA
Professor of Biology
Ball State University
Muncie, IN

Donna L. Ross, PhD
Associate Professor
San Diego State University
San Diego, CA

Marion B. Sewer, PhD
Assistant Professor
School of Biology
Georgia Institute of Technology
Atlanta, GA

Julia Meyer Sheets, PhD
Lecturer
School of Earth Sciences
The Ohio State University
Columbus, OH

Michael J. Singer, PhD
Professor of Soil Science
Department of Land, Air and
Water Resources
University of California
Davis, CA

Karen S. Sottosanti, MA
Science Writer
Pickerington, Ohio

Paul K. Strode, PhD
I.B. Biology Teacher
Fairview High School
Boulder, CO

Jan M. Vermilye, PhD
Research Geologist
Seismo-Tectonic Reservoir
Monitoring (STRM)
Boulder, CO

Judith A. Yero, MA
Director
Teacher's Mind Resources
Hamilton, MT

Dinah Zike, MEd
Author, Consultant,
Inventor of Foldables
Dinah Zike Academy;
Dinah-Might Adventures, LP
San Antonio, TX

Margaret Zorn, MS
Science Writer
Yorktown, VA

Consulting Authors

Alton L. Biggs
Biggs Educational Consulting
Commerce, TX

Ralph M. Feather, Jr., PhD
Assistant Professor
Department of Educational
Studies and Secondary
Education
Bloomsburg University
Bloomsburg, PA

Douglas Fisher, PhD
Professor of Teacher Education
San Diego State University
San Diego, CA

Edward P. Ortleb
Science/Safety Consultant
St. Louis, MO

Series Consultants

Science

Solomon Bililign, PhD
Professor
Department of Physics
North Carolina Agricultural
and Technical State University
Greensboro, NC

John Choinski
Professor
Department of Biology
University of Central Arkansas
Conway, AR

Anastasia Chopelas, PhD
Research Professor
Department of Earth and
Space Sciences
UCLA
Los Angeles, CA

David T. Crowther, PhD
Professor of Science Education
University of Nevada, Reno
Reno, NV

A. John Gatz
Professor of Zoology
Ohio Wesleyan University
Delaware, OH

Sarah Gille, PhD
Professor
University of California
San Diego
La Jolla, CA

David G. Haase, PhD
Professor of Physics
North Carolina State
University
Raleigh, NC

Janet S. Herman, PhD
Professor
Department of Environmental
Sciences
University of Virginia
Charlottesville, VA

David T. Ho, PhD
Associate Professor
Department of Oceanography
University of Hawaii
Honolulu, HI

Ruth Howes, PhD
Professor of Physics
Marquette University
Milwaukee, WI

**Jose Miguel Hurtado, Jr.,
PhD**
Associate Professor
Department of Geological
Sciences
University of Texas at El Paso
El Paso, TX

Monika Kress, PhD
Assistant Professor
San Jose State University
San Jose, CA

Mark E. Lee, PhD
Associate Chair & Assistant
Professor
Department of Biology
Spelman College
Atlanta, GA

Linda Lundgren
Science writer
Lakewood, CO

Carolyn Elliott
Iredell-Statesville Schools
Statesville, NC

Christine M. Jacobs
Ranger Middle School
Murphy, NC

Jason O. L. Johnson
Thurmont Middle School
Thurmont, MD

Felecia Joiner
Stony Point Ninth Grade
Center
Round Rock, TX

Joseph L. Kowalski, MS
Lamar Academy
McAllen, TX

Brian McClain
Amos P. Godby High School
Tallahassee, FL

Von W. Mosser
Thurmont Middle School
Thurmont, MD

Ashlea Peterson
Heritage Intermediate Grade
Center
Coweta, OK

Nicole Lenihan Rhoades
Walkersville Middle School
Walkersvillle, MD

Maria A. Rozenberg
Indian Ridge Middle School
Davie, FL

Barb Seymour
Westridge Middle School
Overland Park, KS

Ginger Shirley
Our Lady of Providence
Junior-Senior High School
Clarksville, IN

Curtis Smith
Elmwood Middle School
Rogers, AR

Sheila Smith
Jackson Public School
Jackson, MS

Sabra Soileau
Moss Bluff Middle School
Lake Charles, LA

Tony Spoores
Switzerland County Middle
School
Vevay, IN

Nancy A. Stearns
Switzerland County Middle
School
Vevay, IN

Kari Vogel
Princeton Middle School
Princeton, MN

Alison Welch
Wm. D. Slider Middle School
El Paso, TX

Linda Workman
Parkway Northeast Middle
School
Creve Coeur, MO

Teacher Advisory Board

The Teacher Advisory Board gave the authors, editorial staff, and design team feedback on the content and design of the Student Edition. They provided valuable input in the development of *Glencoe ⊙Science*.

Frances J. Baldridge
Department Chair
Ferguson Middle School
Beavercreek, OH

Jane E. M. Buckingham
Teacher
Crispus Attucks Medical
Magnet High School
Indianapolis, IN

Elizabeth Falls
Teacher
Blalack Middle School
Carrollton, TX

Nelson Farrier
Teacher
Hamlin Middle School
Springfield, OR

Michelle R. Foster
Department Chair
Wayland Union
Middle School
Wayland, MI

Rebecca Goodell
Teacher
Reedy Creek Middle School
Cary, NC

Mary Gromko
Science Supervisor K–12
Colorado Springs District 11
Colorado Springs, CO

Randy Mousley
Department Chair
Dean Ray Stucky
Middle School
Wichita, KS

David Rodriguez
Teacher
Swift Creek Middle School
Tallahassee, FL

Derek Shook
Teacher
Floyd Middle Magnet School
Montgomery, AL

Karen Stratton
Science Coordinator
Lexington School District One
Lexington, SC

Stephanie Wood
Science Curriculum Specialist,
K–12
Granite School District
Salt Lake City, UT

Online Guide

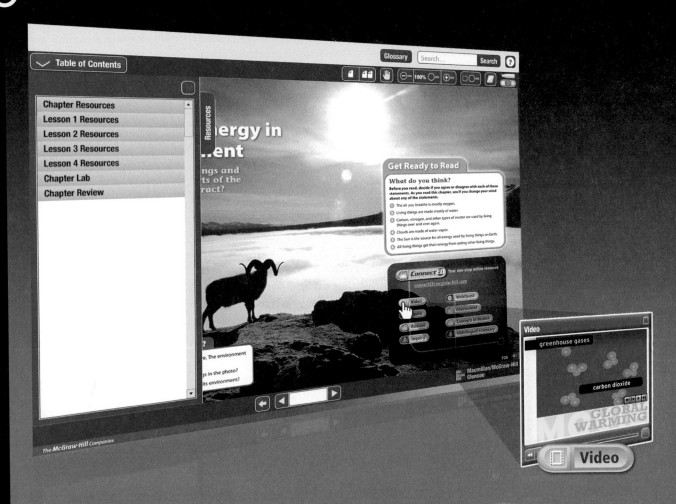

ConnectED

▶ **Your Digital Science Portal**

 **Video**

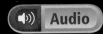 **Audio**

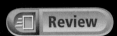 **Review**

 **Inquiry**

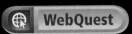 **WebQuest**

See the science in real life through these exciting videos.

Click the link and you can listen to the text while you follow along.

Try these interactive tools to help you review the lesson concepts.

Explore concepts through hands–on and virtual labs.

These web-based challenges relate the concepts you're learning about to the latest news and research.

The icons in your online student edition link you to interactive learning opportunities. Browse your online student book to find more.

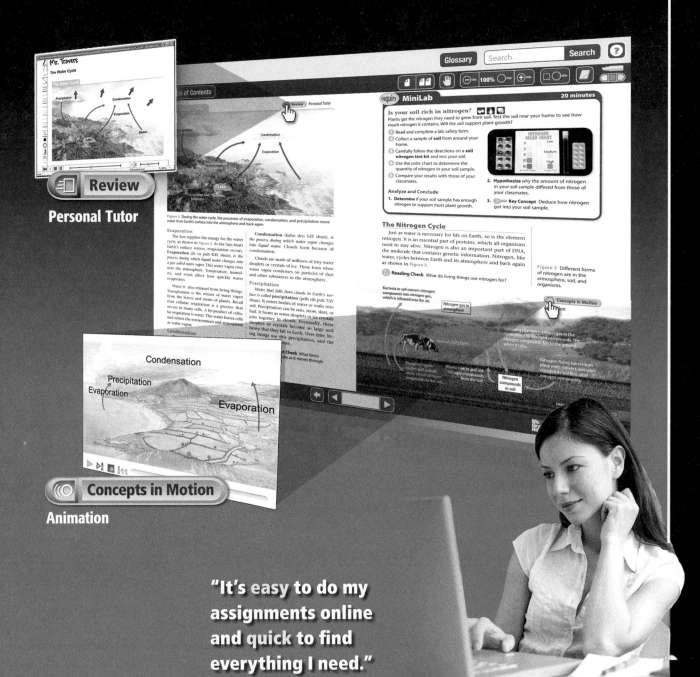

Review

Personal Tutor

Concepts in Motion

Animation

"It's easy to do my assignments online and quick to find everything I need."

Assessment

Check how well you understand the concepts with online quizzes and practice questions.

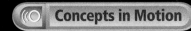

Concepts in Motion

The textbook comes alive with animated explanations of important concepts.

Multilingual eGlossary

Read key vocabulary in 13 languages.

Treasure Hunt

START

Your science book has many features that will aid you in your learning. Some of these features are listed below. You can use the activity at the right to help you find these and other special features in the book.

- **THE BIG IDEA** can be found at the start of each chapter.

- The Reading Guide at the start of each lesson lists 🔑 **Key Concepts,** vocabulary terms, and online supplements to the content.

- 💻 **Connect ED** icons direct you to online resources such as animations, personal tutors, math practices, and quizzes.

- **Inquiry** Labs and Skill Practices are in each chapter.

- Your **FOLDABLES** help organize your notes.

1 What four margin items can help you build your vocabulary?

2 On what page does the glossary begin? What glossary is online?

3 In which Student Resource at the back of your book can you find a listing of Laboratory Safety Symbols?

4 Suppose you want to find a list of all the Launch Labs, MiniLabs, Skill Practices, and Labs, where do you look?

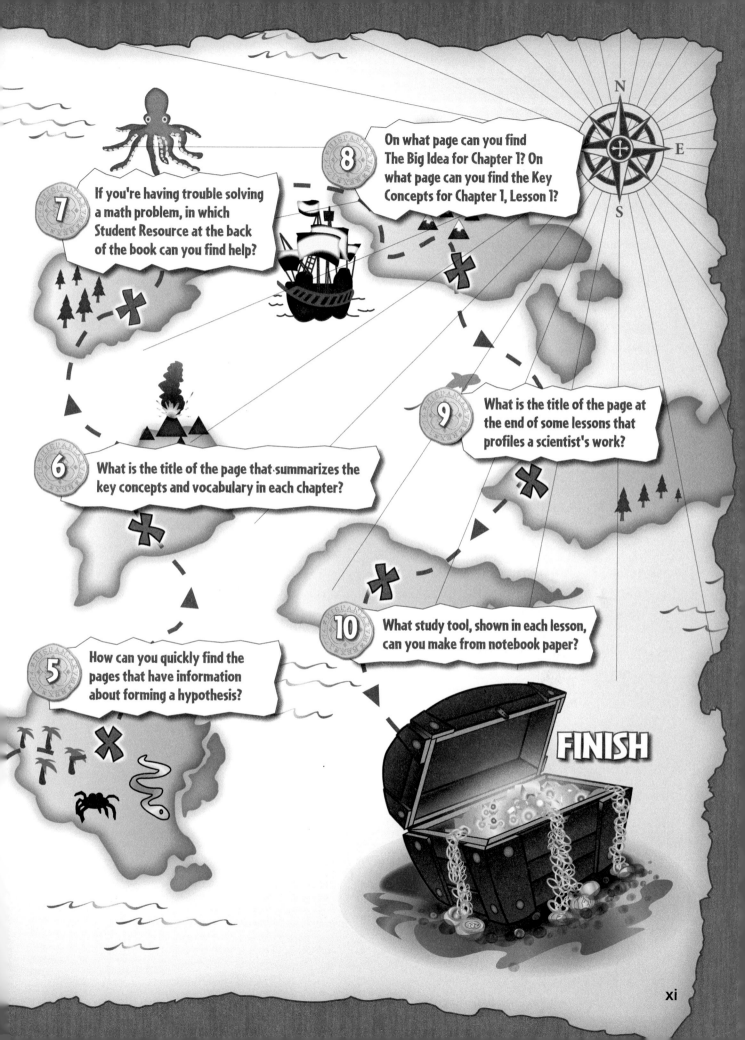

7 If you're having trouble solving a math problem, in which Student Resource at the back of the book can you find help?

8 On what page can you find The Big Idea for Chapter 1? On what page can you find the Key Concepts for Chapter 1, Lesson 1?

9 What is the title of the page at the end of some lessons that profiles a scientist's work?

6 What is the title of the page that summarizes the key concepts and vocabulary in each chapter?

5 How can you quickly find the pages that have information about forming a hypothesis?

10 What study tool, shown in each lesson, can you make from notebook paper?

FINISH

Table of Contents

Table of Contents

Student Resources

TABLE OF CONTENTS

Inquiry

Inquiry Launch Labs

Inquiry MiniLabs

Inquiry Skill Practice

Inquiry Labs

Features

GREEN SCIENCE

SCIENCE & SOCIETY

CAREERS in SCIENCE

INTERACTIONS OF LIFE

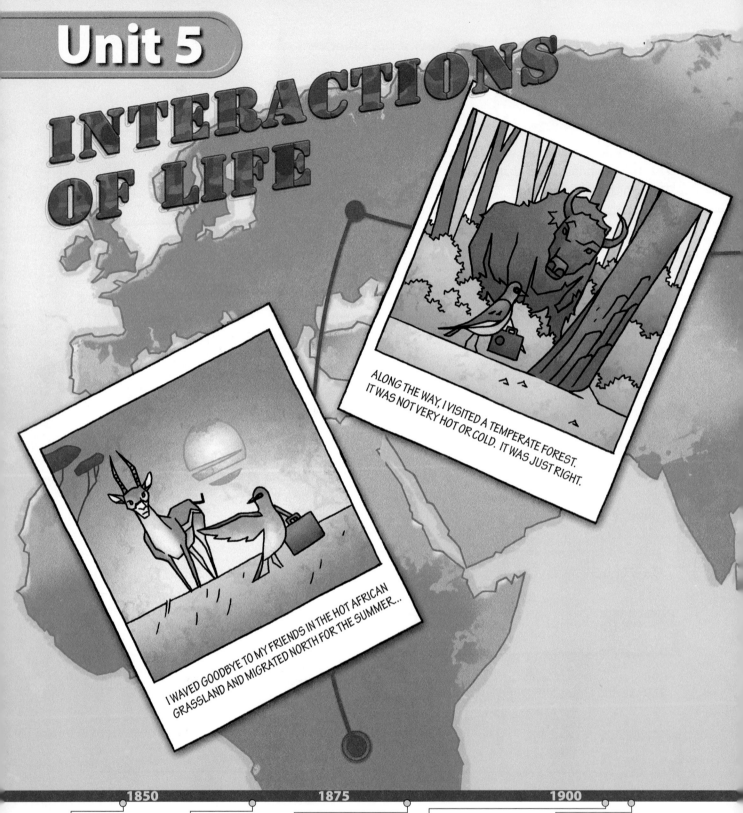

ALONG THE WAY, I VISITED A TEMPERATE FOREST. IT WAS NOT VERY HOT OR COLD. IT WAS JUST RIGHT.

I WAVED GOODBYE TO MY FRIENDS IN THE HOT AFRICAN GRASSLAND AND MIGRATED NORTH FOR THE SUMMER...

1850   **1875**   **1900**

1849
The U.S. Department of Interior is established and is responsible for the management and conservation of most federal land.

1872
The world's first national park, Yellowstone, is created.

1892
The Sierra Club is founded in San Francisco by John Muir. It goes on to be the oldest and largest grassroots environmental organization in the United States.

1915
Congress passes a bill establishing Rocky Mountain National Park in Colorado.

1920
Congress passes the Federal Water Power Act. This act creates a Federal Power Commission with authority over waterways, and the construction and use of water-power projects.

THEN, I RESTED FOR A FEW DAYS IN A COLD, RAINY, MOUNTAIN FOREST IN ASIA. I SAW BEAUTIFUL BAMBOO TREES AND GIANT PANDAS.

I SAID "HI" TO MY FRIENDS IN THE ARCTIC CIRCLE BEFORE HEADING HOME TO THE ALASKAN TUNDRA. NEXT FALL IT'S BACK TO AFRICA FOR THE WINTER.

1950

2000

1955

The Air Pollution Control Act is the first of several United States Clean Air Acts to control air pollution on a national level.

1990

The Clean Air Act Amendments propose emissions trading and add provisions for reducing acid rain, ozone depletion, and toxic air pollution. They also establish a national permits program.

2006

The documentary *An Inconvenient Truth* is released to educate about global warming. The film's popularity raises international awareness of the cause.

? Inquiry

Visit ConnectED for this unit's **STEM** activity.

History and Science

Nearly 50,000 years ago, a group of hunter-gatherers might have roamed through the forest searching for food among the lush plants. The plants and animals that lived in that environment provided the nutritional needs of these people. Humans adapted to the nutrients that the wild foods contained.

Many of the foods you eat today are very different from those eaten by hunter-gatherers. **Table 1** shows how some of these changes occurred.

Table 1 How Science Has Changed Foods

	What?	Advantages	Disadvantages
	Gathering Wild Foods— Foods found in nature were the diet of humans until farming began around 12,000 years ago.	Wild foods provided all the nutrients needed by the human body.	Finding wild foods is not reliable. People moved from place to place in search of food. Sometimes they didn't find food, so they went hungry or starved.
	Farming—People grew seeds from the plants they ate. Tribes settled on land near where they grew crops. If soil conditions were not ideal, farmers learned to add water or animal manure to improve plant growth.	Farming allowed more food to be grown in less space. Over time, people learned to breed plants for larger size or greater disease resistance.	Sometimes, farming changed the nutrient content of foods. People began to suffer from nutrient deficiencies and were prone to disease.
	Hybridizing Plants— Gregor Mendel fertilized one plant with the genetic material from another, producing a hybrid. A hybrid is the offspring of two genetically different organisms.	Hybridization produced new plant foods that combined the best qualities of two plants. The variety of plants available for food increased.	Hybrid crops are prone to disease because of their genetic similarity. Seeds from hybrids do not always grow into plants that produce food of the same quality as the hybrid.
	Genetically Modified (GM) Foods—Scientists remove or replace genes to improve a plant. For example, removing the gene that controls flowering in spinach results in more leaves.	GM plants can increase crop yields, nutrient content, insect resistance, and shelf-life of foods. The lettuce shown here has been modified to produce insulin.	Inserted genes might spread to other plants, producing "superweeds." Allergies to GM foods might increase. The long-term effect on humans is unknown.

A Matter of Taste

In early history, food was eaten raw, just as it was found in nature. Cooking food probably occurred by accident. Someone might have accidentally dropped a root into a fire. When people ate the burnt root, it might have tasted better or been easier to chew. This possibly led to cooking more foods. Over many generations, and with the influence of different cultures and their various ways to prepare food, the taste buds of people changed. People no longer enjoy as many raw foods.

Today, the taste buds of some people tempt them to eat high-calorie, low-nutrition processed foods, as shown in **Figure 1.** These foods contain large amounts of calories, salt, and fat.

In some parts of the world, people buy and prepare fresh fruits and vegetables every day, as shown in Figure 2. In general, these people have lower rates of obesity and fewer diseases that are common in people who eat more processed foods.

One scientist noted that people with a diet very different from our prehistoric ancestors are more susceptible to heart disease, cancer, diabetes, and other "diseases of civilization." Time to take another bite of your fruits and veggies!

Figure 1 Processing foods increases convenience but removes nutrients and adds calories that could lead to obesity.

Figure 2 People in China shop in markets where farmers sell fresh produce that comes directly from the farms.

Inquiry MiniLab
30 minutes

What food would you design?

Suppose you could breed any two fruits or vegetables. What hybrid would you produce?

1. With a partner, discuss qualities of fruits and vegetables that you like and don't like. Then decide on a combination that would have the best qualities of two fruits and/or vegetables.

2. Draw and describe your new food in your Science Journal.

3. Develop a 20-second infomercial advertising the benefits of your new food, and present it to your class.

Analyze and Conclude

1. **Explain** What qualities of the original foods does your hybrid combine?

2. **Predict** Why would people buy your hybrid?

3. **Explain** What are some advantages and disadvantages of your hybrid?

Chapter 20

Matter and Energy in the Environment

THE BIG IDEA How do living things and the nonliving parts of the environment interact?

Inquiry   How does the ram survive?

The ram needs food, air, water, and shelter to survive. The environment provides the ram with all that it needs to survive.

- How does the ram depend on the nonliving things in the photo?

- How might the ram interact with living things in its environment?

Get Ready to Read

What do you think?

Before you read, decide if you agree or disagree with each of these statements. As you read this chapter, see if you change your mind about any of the statements.

1 The air you breathe is mostly oxygen.

2 Living things are made mostly of water.

3 Carbon, nitrogen, and other types of matter are used by living things over and over again.

4 Clouds are made of water vapor.

5 The Sun is the source for all energy used by living things on Earth.

6 All living things get their energy from eating other living things.

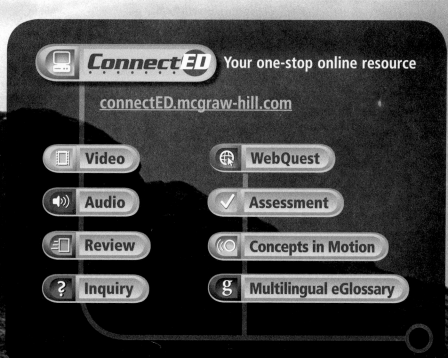

ConnectED Your one-stop online resource

connectED.mcgraw-hill.com

- Video
- Audio
- Review
- Inquiry
- WebQuest
- Assessment
- Concepts in Motion
- Multilingual eGlossary

Abiotic Factors

Reading Guide

Key Concepts

ESSENTIAL QUESTIONS

- What are the nonliving parts of an environment?

Vocabulary

ecosystem p. 707

biotic factor p. 707

abiotic factor p. 707

climate p. 708

atmosphere p. 709

 g **Multilingual eGlossary**

Video **BrainPOP®**

Inquiry Why So Blue?

Have you ever seen a picture of a bright blue ocean? The water looks so colorful in part because of nonliving factors such as matter in the water and the gases surrounding Earth. These nonliving things change the way you see light from the Sun, another nonliving part of the environment.

Is it living or nonliving?

You are surrounded by living and nonliving things, but it is sometimes difficult to tell what is alive. Some nonliving things may appear to be alive at first glance. Others are alive or were once living, but seem nonliving. In this lab, you will explore which items are alive and which are not.

1 Draw a chart with the headings *Living* and *Nonliving*.

2 Your teacher will provide you with a list of items. Decide if each item is living or nonliving.

Living	Nonliving

Think About This

1. What are some characteristics that the items in the *Living* column share?

2. **Key Concept** How might the nonliving items be a part of your environment?

What is an ecosystem?

Have you ever watched a bee fly from flower to flower? Certain flowers and bees depend on each other. Bees help flowering plants reproduce. In return, flowers provide the nectar that bees use to make honey. Flowers also need nonliving things to survive, such as sunlight and water. For example, if plants don't get enough water, they can die. The bees might die, too, because they feed on the plants. All organisms need both living and nonliving things to survive.

An **ecosystem** *is all the living things and nonliving things in a given area.* Ecosystems vary in size. An entire forest can be an ecosystem, and so can a rotting log on the forest floor. Other examples of ecosystems include a pond, a desert, an ocean, and your neighborhood.

Biotic (bi AH tihk) **factors** *are the living things in an eco-system.* **Abiotic** (ay bi AH tihk) **factors** *are the nonliving things in an ecosystem, such as sunlight and water.* Biotic factors and abiotic factors depend on each other. If just one factor—either abiotic or biotic—is disturbed, other parts of the ecosystem are affected. For example, severe droughts, or periods of water shortages, occurred in Australia in 2006. Many fish in rivers and lakes died. Animals that fed on the fish had to find food elsewhere. A lack of water, an abiotic factor, affected biotic factors in this ecosystem, such as the fish and the animals that fed on the fish.

WORD ORIGIN

biotic
from Greek *biotikos*, means "fit for life"

Figure 1 Abiotic factors include sunlight, water, atmosphere, soil, temperature, and climate.

What are the nonliving parts of an ecosystem?

Some abiotic factors in an ecosystem are shown in **Figure 1**. Think about how these factors might affect you. You need sunlight for warmth and air to breathe. You would have no food without water and soil. These nonliving parts of the environment affect all living things.

The Sun

The source of almost all energy on Earth is the Sun. It provides warmth and light. In addition, many plants use sunlight and make food, as you'll read in Lesson 3. The Sun also affects two other abiotic factors—climate and temperature.

Reading Check How do living things use the Sun's energy?

Climate

Polar bears live in the Arctic. The Arctic has a cold, dry climate. **Climate** *describes average weather conditions in an area over time.* These weather conditions include temperature, moisture, and wind.

Climate influences where organisms can live. A desert climate, for example, is dry and often hot. A plant that needs a lot of water could not survive in a desert. In contrast, a cactus is well adapted to a dry climate because it can survive with little water.

Temperature

Is it hot or cold where you live? Temperatures on Earth vary greatly. Temperature is another abiotic factor that influences where organisms can survive. Some organisms, such as tropical birds, thrive in hot conditions. Others, such as polar bears, are well adapted to the cold. Tropical birds don't live in cold ecosystems, and polar bears don't live in warm ecosystems.

Water

All life on Earth requires water. In fact, most organisms are made mostly of water. All organisms need water for important life processes, such as growing and reproducing. Every ecosystem must contain some water to support life.

Fungi

Gases in Atmosphere

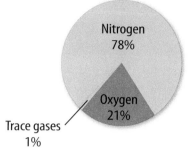

Nitrogen 78%

Oxygen 21%

Trace gases 1%

Climate

(graph showing Precipitation (cm) on left axis 0–30 and Temperature (°C) on right axis 0–30, by Month from Jan to Nov, with Temperature line and Precipitation bars)

Temperature
Precipitation

Precipitation (cm)

Temperature (°C)

Month

Jan   Mar   May   Jul   Sep   Nov

Visual Check How does the jaguar interact with the abiotic factors in its ecosystem?

Atmosphere

Every time you take a breath you are interacting with another abiotic factor that is necessary for life—the atmosphere. *The* **atmosphere** *(AT muh sfir) is the layer of gases that surrounds Earth.* The atmosphere is mostly nitrogen and oxygen with trace amounts of other gases, also shown in **Figure 1.** Besides providing living things with oxygen, the atmosphere also protects them from certain harmful rays from the Sun.

Soil

Bits of rocks, water, air, minerals, and the remains of once-living things make up soil. When you think about soil, you might picture a farmer growing crops. Soil provides water and nutrients for the plants we eat. However, it is also a home for many organisms, such as insects, bacteria, and fungi.

Factors such as water, soil texture, and the amount of available nutrients affect the types of organisms that can live in soil. Bacteria break down dead plants and animals, returning nutrients to the soil. Earthworms and insects make small tunnels in the soil, allowing air and water to move through it. Even very dry soil, like that in the desert, is home to living things.

 Key Concept Check List the nonliving things in ecosystems.

FOLDABLES

Fold and cut a sheet of paper to make a six-door book. Label it as shown. Use it to organize information about the abiotic parts of an ecosystem.

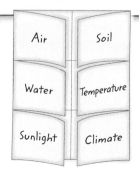

Air | Soil
Water | Temperature
Sunlight | Climate

Lesson 1 Review

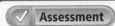

Visual Summary

Ecosystems include all the biotic and abiotic factors in an area.

Biotic factors are the living things in ecosystems.

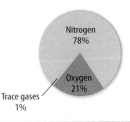

Abiotic factors are the nonliving things in ecosystems, including water, sunlight, temperature, climate, air, and soil.

FOLDABLES®

Use your lesson Foldable to review the lesson. Save your Foldable for the project at the end of the chapter.

Use Vocabulary

1. **Distinguish** between biotic and abiotic factors.

2. **Define** *ecosystem* in your own words.

3. **Use the term** *climate* in a complete sentence.

Understand Key Concepts

4. What role do bacteria play in soil ecosystems?
   - A. They add air to soil.
   - B. They break down rocks.
   - C. They return nutrients to soil.
   - D. They tunnel through soil.

5. **Explain** How would a forest ecosystem change if no sunlight were available to it?

Interpret Graphics

6. **Analyze** The graph below shows climate data for an area. How would you describe this climate?

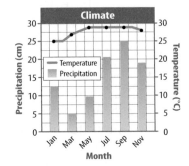

7. **Organize** Copy the graphic organizer below and fill in each oval with an abiotic factor.

Critical Thinking

8. **Predict** Imagine that the soil in an area is carried away by wind and water, leaving only rocks behind. How would this affect the living things in that area?

What do you think NOW?

You first read the statements below at the beginning of the chapter.

1. The air you breathe is mostly oxygen.

2. Living things are made mostly of water.

Did you change your mind about whether you agree or disagree with the statements? Rewrite any false statements to make them true.

Terraforming Mars

Life on Another Planet

Terraforming is the process of transforming an environment that cannot support life into one that can. Making Mars like Earth would take more than just growing plants and adding water. You would need to consider how every abiotic factor needed to support life would be included in the new environment.

First, consider Mars's temperature. Although Mars gets plenty of sunlight, it is farther from the Sun than Earth is. Air temperatures go no higher than 0°C on a midsummer Martian day. Don't even think about trying to survive a winter night on Mars, as temperatures fall below −89°C.

How could you change the temperature on Mars? Releasing greenhouse gases such as chlorofluorocarbons (CFCs) into the atmosphere can cause the planet to get warmer. Raising the average temperature by only 4°C would melt the polar ice caps, releasing frozen CO_2, another greenhouse gas. This also would cause bodies of water to form. As temperatures rise, liquid water trapped in the soil would turn into a gas, providing the planet with water vapor, an important abiotic factor.

With water and warmer temperatures, plant life could be introduced. While turning light energy into food, plants would introduce another abiotic factor—oxygen. With all the needed abiotic factors accounted for, NASA scientists think that in a few centuries Mars could support life similar to that on Earth.

SCIENCE & SOCIETY

◀ Mars is cold and dry, with no sign of life on its dusty, red surface.

Life as it is on Earth does not exist on Mars. However, when you compare all the planets in our solar system, Mars is the most like Earth.

It's Your Turn

DEBATE Why would people want to move to Mars? Would this be the right choice? Research these questions and then debate the issues.

Reading Guide

Key Concepts
ESSENTIAL QUESTIONS

- How does matter move in ecosystems?

Vocabulary

evaporation p. 714

condensation p. 714

precipitation p. 714

nitrogen fixation p. 716

g Multilingual eGlossary

Video **BrainPOP®**

Cycles of Matter

Inquiry **Where does the water go?**

All water, including the water in this waterfall, can move throughout an ecosystem in a cycle. It can also change forms. What other forms do you think water takes as it moves through an ecosystem?

How can you model raindrops?

Like all matter on Earth, water is recycled. It constantly moves between Earth and its atmosphere. You could be drinking the same water that a *Tyrannosaurus rex* drank 65 million years ago!

1. Read and complete a lab safety form.
2. Half-fill a **plastic cup** with warm water.
3. Cover the cup with **plastic wrap.** Secure the plastic with a **rubber band.**
4. Place an **ice cube** on the plastic wrap. Observe the cup for several minutes. Record your observations in your Science Journal.

Think About This

1. What did you observe on the underside of the plastic wrap? Why do you think this happened?

2. How does this activity model the formation of raindrops?

3. 🔑 **Key Concept** Do you think other matter moves through the environment? Explain your answer.

How does matter move in ecosystems?

The water that you used to wash your hands this morning might have once traveled through the roots of a tree in Africa or even have been part of an Antarctic glacier. How can this be? Water moves continuously through ecosystems. It is used over and over again. The same is true of carbon, oxygen, nitrogen, and other types of matter. Elements that move through one matter cycle may also play a role in another, such as oxygen's role in the water cycle.

The Water Cycle

Look at a globe or a map. Notice that water surrounds the landmasses. Water covers about 70 percent of Earth's surface.

Most of Earth's water—about 97 percent—is in oceans. Water is also in rivers and streams, lakes, and underground reservoirs. In addition, water is in the atmosphere, icy glaciers, and living things.

Water continually cycles from Earth to its atmosphere and back again. This movement of water is called the water cycle. It involves three processes: evaporation, condensation, and precipitation.

SCIENCE USE V. COMMON USE

element
Science Use one of a class of substances that cannot be separated into simpler substances by chemical means

Common Use a part or piece

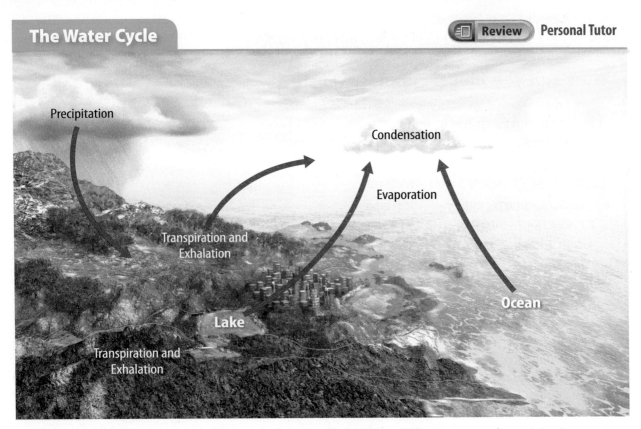

Figure 2 During the water cycle, the processes of evaporation, condensation, and precipitation move water from Earth's surface into the atmosphere and back again.

Evaporation

The Sun supplies the energy for the water cycle, as shown in **Figure 2.** As the Sun heats Earth's surface waters, evaporation occurs. **Evaporation** (ih va puh RAY shun), *is the process during which liquid water changes into a gas called water vapor.* This water vapor rises into the atmosphere. Temperature, humidity, and wind affect how quickly water evaporates.

Water is also released from living things. Transpiration is the release of water vapor from the leaves and stems of plants. Recall that cellular respiration is a process that occurs in many cells. A by-product of cellular respiration is water. This water leaves cells and enters the environment and atmosphere as water vapor.

Condensation

The higher in the atmosphere you are, the cooler the temperature is. As water vapor rises, it cools and condensation occurs.

Condensation (kahn den SAY shun), *is the process during which water vapor changes into liquid water.* Clouds form because of condensation.

Clouds are made of millions of tiny water droplets or crystals of ice. These form when water vapor condenses on particles of dust and other substances in the atmosphere.

Precipitation

Water that falls from clouds to Earth's surface is called **precipitation** (prih sih puh TAY shun). It enters bodies of water or soaks into soil. Precipitation can be rain, snow, sleet, or hail. It forms as water droplets or ice crystals join together in clouds. Eventually, these droplets or crystals become so large and heavy that they fall to Earth. Over time, living things use this precipitation, and the water cycle continues.

 Key Concept Check What forms does water take as it moves through ecosystems?

Is your soil rich in nitrogen?

Plants get the nitrogen they need to grow from soil. Test the soil near your home to see how much nitrogen it contains. Will the soil support plant growth?

1. Read and complete a lab safety form.
2. Collect a sample of **soil** from around your home.
3. Carefully follow the directions on a **soil nitrogen test kit** and test your soil.
4. Use the color chart to determine the quantity of nitrogen in your soil sample.
5. Compare your results with those of your classmates.

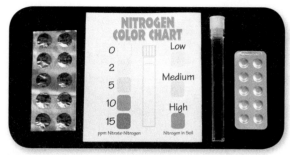

Analyze and Conclude

1. **Determine** if your soil sample has enough nitrogen to support most plant growth.

2. **Hypothesize** why the amount of nitrogen in your soil sample differed from those of your classmates.

3. 🔑 **Key Concept** Deduce how nitrogen got into your soil sample.

The Nitrogen Cycle

Just as water is necessary for life on Earth, so is the element nitrogen. It is an essential part of proteins, which all organisms need to stay alive. Nitrogen is also an important part of DNA, the molecule that contains genetic information. Nitrogen, like water, cycles between Earth and its atmosphere and back again as shown in **Figure 3**.

 Reading Check What do living things use nitrogen for?

Figure 3 Different forms of nitrogen are in the atmosphere, soil, and organisms.

Bacteria in soil convert nitrogen compounds into nitrogen gas, which is released into the air.

Nitrogen gas in atmosphere

Concepts in Motion

Animation

Lightning changes nitrogen gas in the atmosphere to nitrogen compounds. The nitrogen compounds fall to the ground when it rains.

Animals eat plants.

Nitrogen-fixing bacteria on plant roots convert unusable nitrogen in soil to usable nitrogen compounds.

Decaying organic matter and animal waste return nitrogen compounds to the soil.

Plants take in and use nitrogen compounds from the soil.

Nitrogen compounds in soil

From the Environment to Organisms

Recall that the atmosphere is mostly nitrogen. However, this nitrogen is in a form that plants and animals cannot use. How do organisms get nitrogen into their bodies? The nitrogen must first be changed into a different form with the help of certain bacteria that live in soil and water. These bacteria take in nitrogen from the atmosphere and change it into nitrogen compounds that other living things can use. *The process that changes atmospheric nitrogen into nitrogen compounds that are usable by living things is called* **nitrogen fixation** (NI truh jun • fihk SAY shun). Nitrogen fixation is shown in **Figure 4.**

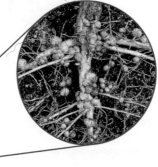

◀ **Figure 4** Certain bacteria convert nitrogen in soil and water into a form usable by plants.

Plants and some other organisms take in this changed nitrogen from the soil and water. Then, animals take in nitrogen when they eat the plants or other organisms.

✓ **Reading Check** What is nitrogen fixation?

From Organisms to the Environment

Some types of bacteria can break down the tissues of dead organisms. When organisms die, these bacteria help return the nitrogen in the tissues of dead organisms to the environment. This process is shown in **Figure 5.**

Nitrogen also returns to the environment in the waste products of organisms. Farmers often spread animal wastes, called manure, on their fields during the growing season. The manure provides nitrogen to plants for better growth.

◀ **Figure 5** Bacteria break down the remains of dead plants and animals.

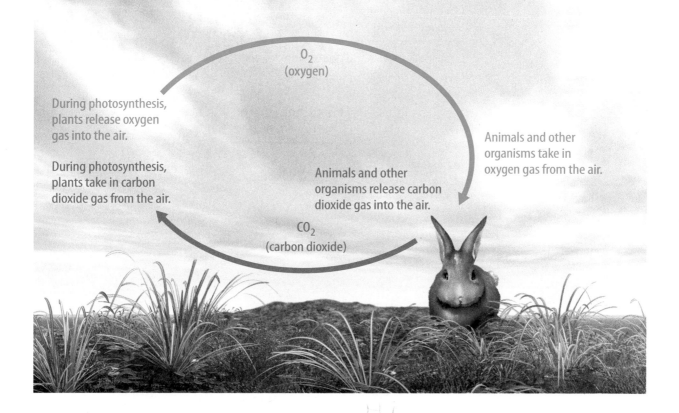

O₂
(oxygen)

During photosynthesis, plants release oxygen gas into the air.

During photosynthesis, plants take in carbon dioxide gas from the air.

Animals and other organisms release carbon dioxide gas into the air.

Animals and other organisms take in oxygen gas from the air.

CO₂
(carbon dioxide)

The Oxygen Cycle

Almost all living things need oxygen for cellular processes that release energy. Oxygen is also part of many substances that are important to life, such as carbon dioxide and water. Oxygen cycles through ecosystems, as shown in **Figure 6.**

Earth's early atmosphere probably did not contain oxygen gas. Oxygen might have entered the atmosphere when certain bacteria evolved that could carry out the process of photosynthesis and make their own food. A by-product of photosynthesis is oxygen gas. Over time, other photosynthetic organisms evolved and the amount of oxygen in Earth's atmosphere increased. Today, photosynthesis is the primary source of oxygen in Earth's atmosphere. Some scientists estimate that unicellular organisms in water, called phytoplankton, release more than 50 percent of the oxygen in Earth's atmosphere.

Many living things, including humans, take in the oxygen and release carbon dioxide. The interaction of the carbon and oxygen cycles is one example of a relationship between different types of matter in ecosystems. As the matter cycles through an ecosystem, both the carbon and the oxygen take different forms and play a role in the other element's cycle.

Figure 6 Most oxygen in the air comes from plants and algae.

Visual Check Describe your part in the oxygen cycle.

REVIEW VOCABULARY

bacteria
a group of microscopic unicellular organisms without a membrane-bound nucleus

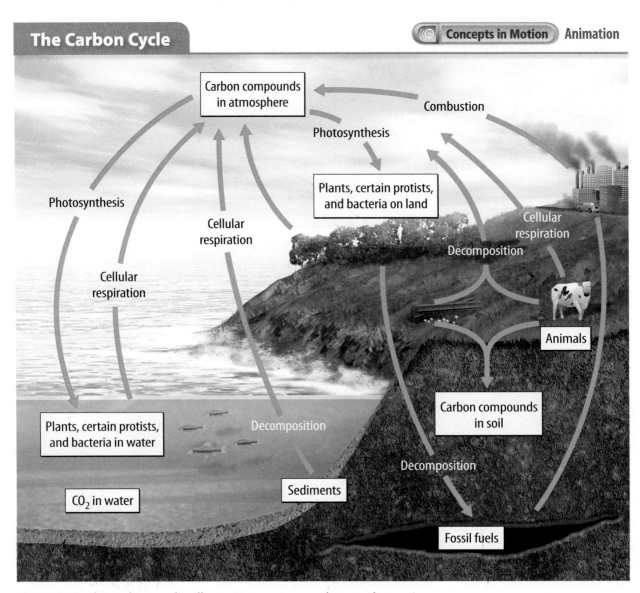

Figure 7 In the carbon cycle, all organisms return carbon to the environment.

The Carbon Cycle

All organisms contain carbon. It is part of proteins, sugars, fats, and DNA. Some organisms, including humans, get carbon from food. Other organisms, such as plants, get carbon from the atmosphere or bodies of water. Like other types of matter, carbon cycles through ecosystems, as shown in **Figure 7.**

Carbon in Soil

Like nitrogen, carbon can enter the environment when organisms die and decompose. This returns carbon compounds to the soil and releases carbon dioxide (CO_2) into the atmosphere for use by other organisms. Carbon is also found in fossil fuels, which formed when decomposing organisms were exposed to pressure, high temperatures, and bacteria over hundreds of millions of years.

FOLDABLES

Make a half book from a sheet of paper. Select a cycle of matter and use your book to organize information about the biotic and abiotic parts of that cycle.

Cycles in Nature

Carbon in Air

Recall that carbon is found in the atmosphere as carbon dioxide. Plants and other photosynthetic organisms take in carbon dioxide and water and produce energy-rich sugars. These sugars are a source of carbon and energy for organisms that eat photosynthetic organisms. When the sugar is broken down by cells and its energy is released, carbon dioxide is released as a by-product. This carbon dioxide gas enters the atmosphere, where it can be used again.

The Greenhouse Effect

Carbon dioxide is one of the gases in the atmosphere that absorbs thermal energy from the Sun and keeps Earth warm. This process is called the greenhouse effect. The Sun produces solar radiation, as shown in **Figure 8.** Some of this energy is reflected back into space, and some passes through Earth's atmosphere. Greenhouse gases in Earth's atmosphere absorb thermal energy that reflects off Earth's surface. The more greenhouse gases released, the greater the gas layer becomes and the more thermal energy is absorbed. These gases are one factor that keeps Earth from becoming too hot or too cold.

Reading Check What is the greenhouse effect?

While the greenhouse effect is essential for life, a steady increase in greenhouse gases can harm ecosystems. For example, carbon is stored in fossil fuels such as coal, oil, and natural gas. When people burn fossil fuels to heat homes, for transportation, or to provide electricity, carbon dioxide gas is released into the atmosphere. The amount of carbon dioxide in the air has increased due to both natural and human activities.

ACADEMIC VOCABULARY

release
(verb) to set free or let go

Figure 8 Some thermal energy remains close to the Earth due to greenhouse gases.

Visual Check What might happen if heat were not absorbed by greenhouse gases?

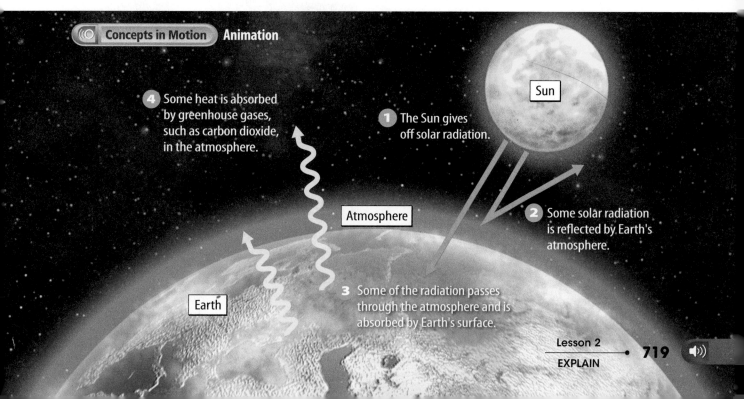

Concepts in Motion Animation

4 Some heat is absorbed by greenhouse gases, such as carbon dioxide, in the atmosphere.

1 The Sun gives off solar radiation.

Sun

Atmosphere

2 Some solar radiation is reflected by Earth's atmosphere.

Earth

3 Some of the radiation passes through the atmosphere and is absorbed by Earth's surface.

Lesson 2 Review

Visual Summary

Matter such as water, oxygen, nitrogen, and carbon cycles through ecosystems.

The three stages of the water cycle are evaporation, condensation, and precipitation.

The greenhouse effect helps keep the Earth from getting too hot or too cold.

FOLDABLES

Use your lesson Foldable to review the lesson. Save your Foldable for the project at the end of the chapter.

What do you think NOW?

You first read the statements below at the beginning of the chapter.

3. Carbon, nitrogen, and other types of matter are used by living things over and over again.

4. Clouds are made of water vapor.

Did you change your mind about whether you agree or disagree with the statements? Rewrite any false statements to make them true.

Use Vocabulary

1 **Distinguish** between evaporation and condensation.

2 **Define** *nitrogen fixation* in your own words.

3 Water that falls from clouds to Earth's surface is called _____.

Understand Key Concepts

4 What is the driving force behind the water cycle?
  A. gravity   C. sunlight
  B. plants   D. wind

5 **Infer** Farmers add nitrogen to their fields every year to help their crops grow. Why must farmers continually add nitrogen when this element recycles naturally?

Interpret Graphics

6 **Sequence** Draw a graphic organizer like the one below and sequence the steps in the water cycle.

7 **Summarize** how the greenhouse effect moderates temperatures on Earth.

Critical Thinking

8 **Explain** how oxygen cycles through the ecosystem in which you live.

9 **Consider** How might ecosystems be affected if levels of atmospheric CO_2 continue to rise?

How do scientists use variables?

Materials

rubber ball

styrene ball

meterstick

Safety

If you wanted to find out what made one ball bounce higher than another, you might design an experiment that uses variables. You could test whether balls made of one material bounce higher than those made of another. By changing only one variable, the experiment tests only the effect of that factor.

Learn It

When experimenting, scientists often **use variables.** A variable is anything that can be changed. For example, a scientist might want to study the effect that different amounts of water have on a plant's growth. The amount of water is the variable in the experiment. Other factors, such as soil type and amount of sunlight, stay the same.

Try It

1. Read and complete a lab safety form.

2. Examine both the rubber ball and the styrene ball. Predict which ball will bounce higher. Record your prediction in your Science Journal.

3. With your partner, hold the rubber ball 35 cm above the table and drop it. Record how high it

bounces. Drop the ball a total of three times, recording the height it bounces each time. Calculate the average height that the rubber ball bounced.

4. Repeat step 3 with the styrene ball.

Apply It

5. Compare the average height of each ball's bounce and determine which bounced higher. Did your data support your prediction?

6. Identify two other variables you could test in this experiment. Would you test them together or separately? Explain.

7. 🔑 **Key Concept** What variables might affect a study of the water cycle in your neighborhood?

	Rubber Ball	Styrene Ball
Trial 1		
Trial 2		
Trial 3		
Average bounce		

Reading Guide

Key Concepts
ESSENTIAL QUESTIONS

- How does energy move in ecosystems?
- How is the movement of energy in an ecosystem modeled?

Vocabulary

photosynthesis p. 724

chemosynthesis p. 724

food chain p. 726

food web p. 727

energy pyramid p. 728

 Multilingual eGlossary

Video **BrainPOP®**

Energy in Ecosystems

Inquiry **Time for a snack?**

All organisms need energy, and many get it from eating other organisms. Can you guess how each of the living things in this picture gets the energy it needs?

How does energy change form?

Every day, sunlight travels hundreds of millions of kilometers and brings warmth and light to Earth. Energy from the Sun is necessary for nearly all life on Earth. Without it, most life could not exist.

1. Read and complete a lab safety form.

2. Obtain **UV-sensitive beads** from your teacher. Write a description of them in your Science Journal.

3. Place half the beads in a sunny place. Place the other half in a dark place.

4. Wait a few minutes, and then observe both sets of beads. Record your observations in your Science Journal.

Think About This

1. Compare and contrast the two sets of beads after the few minutes. How are they different? How are they the same?

2. Hypothesize why the beads looked different.

3. 🔑 **Key Concept** How do you think living things use energy?

How does energy move in ecosystems?

When you see a picture of an ecosystem, it often looks quiet and peaceful. However, ecosystems are actually full of movement. Birds squawk and beat their wings, plants sway in the breeze, and insects buzz.

Each movement made by a living thing requires energy. All of life's functions, including growth and reproduction, require energy. The main source of energy for most life on Earth is the Sun. Unlike other resources, such as water and carbon, energy does not cycle through ecosystems. Instead, energy flows in one direction, as shown in **Figure 9.** In most cases, energy flow begins with the Sun, and moves from one organism to another. Many organisms get energy by eating other organisms. Sometimes organisms change energy into different forms as it moves through an ecosystem. Not all the energy an organism gets is used for life processes. Some is released to the environment as thermal energy. You might have read that energy cannot be created or destroyed, but it can change form. This idea is called the law of conservation of energy.

🔑 **Key Concept Check** How do the movements of matter and energy differ?

Cycle and Flow

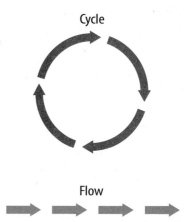

Cycle

Flow

Figure 9 Matter moves in a cycle pattern, and energy moves in a flow pattern.

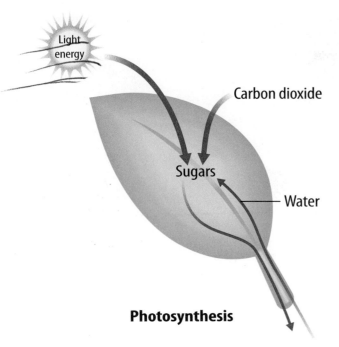

Photosynthesis

▲ **Figure 10** Most producers make their food through the process of photosynthesis.

WORD ORIGIN · · · · · · · · · · · · · · · · · · ·
photosynthesis
from Greek *photo*, meaning "light"; and *synthese*, meaning "synthesis"

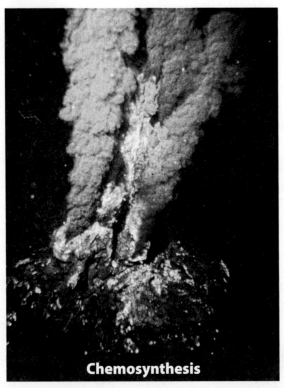

Chemosynthesis

▲ **Figure 11** The producers at a hydrothermal vent make their food using chemosynthesis.

Producers

People who make things or products are often called producers. In a similar way, living things that make their own food are called producers. Producers make their food from materials found in their environments. Most producers are photosynthetic (foh toh sihn THEH tihk). They use the process of photosynthesis (foh toh SIHN thuh sus), which is described below. Grasses, trees, and other plants, algae and some other protists, and certain bacteria are photosynthetic. Other producers, including some bacteria, are chemosynthetic (kee moh sihn THEH tihk). They make their food using chemosynthesis (kee moh SIHN thuh sus).

Photosynthesis Recall that in the carbon cycle, carbon in the atmosphere cycles through producers such as plants, into other organisms, and then back into the atmosphere. This and other matter cycles involve photosynthesis, as shown in **Figure 10.** **Photosynthesis** *is a series of chemical reactions that convert light energy, water, and carbon dioxide into the food-energy molecule glucose and give off oxygen.*

Chemosynthesis As you read earlier, some producers make food using chemosynthesis. **Chemosynthesis** *is the process during which producers use chemical energy in matter rather than light energy and make food.* One place where chemosynthesis can occur is on the deep ocean floor. There, inorganic compounds that contain hydrogen and sulfur, along with thermal energy from Earth's interior, flow out from cracks in the ocean floor. These cracks are called hydrothermal vents. These vents, such as the one shown in **Figure 11,** are home to chemosynthetic bacteria. These bacteria use the chemical energy contained in inorganic compounds in the hot water and produce food.

 Reading Check What materials do producers use to make food during chemosynthesis?

Herbivore

Carnivore

Omnivore

Detritivore

Detritivore—
Decomposer

Figure 12 Organisms can be classified by the type of food that they eat.

Consumers

Some consumers are shown in **Figure 12.** Consumers do not produce their own energy-rich food, as producers do. Instead, they get the energy they need to survive by consuming other organisms.

Consumers can be classified by the type of food that they eat. Herbivores feed on only producers. For example, a deer is an herbivore because it eats only plants. Carnivores eat other animals. They are usually predators, such as lions and wolves. Omnivores eat both producers and other consumers. A bird that eats berries and insects is an omnivore.

Another group of consumers is detritivores (dih TRI tuh vorz). They get their energy by eating the remains of other organisms. Some detritivores, such as insects, eat dead organisms. Other detritivores, such as bacteria and mushrooms, feed on dead organisms and help decompose them. For this reason, these organisms often are called decomposers. During decomposition, decomposers produce carbon dioxide that enters the atmosphere. Some of the decayed matter enters the soil. In this way, detritivores help recycle nutrients through ecosystems. They also help keep ecosystems clean. Without decomposers, dead organisms would pile up in an ecosystem.

Inquiry **MiniLab**   15 minutes

How can you classify organisms?

Most organisms get their energy from the Sun or by consuming other organisms. In this lab, you will use photographs to classify organisms based on their feeding relationships.

1. Search through a **magazine** for photographs of ecosystems. Select a photograph that shows several different organisms.

2. In your Science Journal, identify the organisms. Include a description of their environment.

3. Describe how the organisms are interacting with one another and with their environment.

Analyze and Conclude

1. **Classify** the organisms as producers or consumers. Use details from the photograph to support your classification scheme.

2. 🔑 **Key Concept** How do the producers in your photo obtain energy? How do the consumers get their energy?

Modeling Energy in Ecosystems

Unlike matter, energy does not cycle through ecosystems because it does not return to the Sun. Instead, energy flows through ecosystems. Organisms use some energy for life processes. In addition, organisms store some energy in their bodies as chemical energy. When consumers eat these organisms, this chemical energy moves into the bodies of consumers. However, with each transfer of energy from organism to organism, some energy changes to thermal energy. The bodies of consumers emit excess thermal energy, which then enters the environment. Scientists use models to study this flow of energy through an ecosystem. They use different models depending on how many organisms they are studying.

Food Chains

A **food chain** *is a model that shows how energy flows in an ecosystem through feeding relationships.* In a food chain, arrows show the transfer of energy. A typical food chain is shown in **Figure 13.** Notice that there are not many links in this food chain. That is because the amount of available energy decreases every time it is transferred from one organism to another.

 Key Concept Check How does a food chain model energy flow?

Figure 13 Energy moves from the Sun to a plant, a mouse, a snake, and a hawk in this food chain.

Food Chain

1 The Sun emits energy.

5 The hawk obtains energy by eating the snake.

2 Plants make energy-rich food using sunlight.

3 The mouse obtains energy by eating the plant.

4 The snake obtains energy by eating the mouse.

Food Webs

Imagine you have a jigsaw puzzle of a tropical rain forest. Each piece of the puzzle shows only one small part of the forest. A food chain is like one piece of an ecosystem jigsaw puzzle. It is helpful when studying certain parts of an ecosystem, but it does not show the whole picture.

In the food chain on the previous page, the mouse might also eat the seeds of several producers, such as corn, berries, or grass. The snake might eat other organisms such as frogs, crickets, lizards, or earthworms too.

The hawk hunts mice, squirrels, rabbits, and fish, as well as snakes. *Scientists use a model of energy transfer called a* **food web** *to show how food chains in a community are interconnected,* as shown in **Figure 14.** You can think of a food web as many overlapping food chains. Like in a food chain, arrows show how energy flows in a food web. Some organisms in the food web might be part of more than one food chain in that web.

Key Concept Check What models show the transfer of energy in an ecosystem?

Figure 14 A food web shows the complex feeding relationships among organisms in an ecosystem.

Review   Personal Tutor

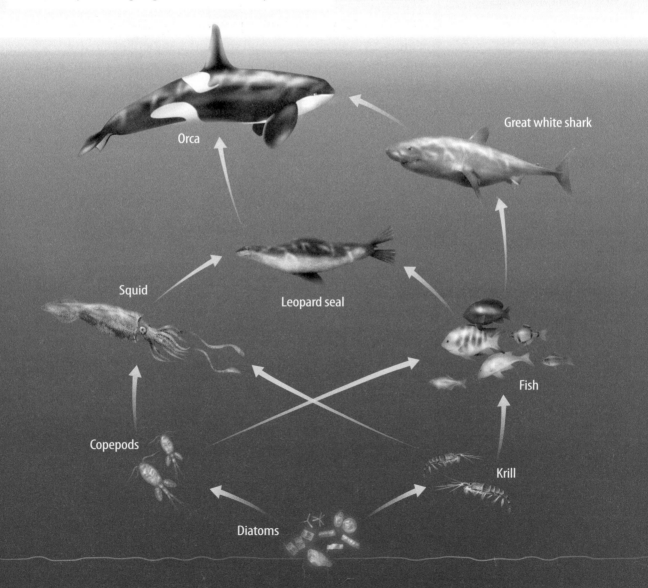

Energy Pyramids

Food chains and food webs show how energy moves in an ecosystem. However, they do not show how the amount of energy in an ecosystem changes. *Scientists use a model called an* **energy pyramid** *to show the amount of energy available in each step of a food chain,* as shown in **Figure 15.** The steps of an energy pyramid are also called trophic (TROH fihk) levels.

Producers, such as plants, make up the trophic level at the bottom of the pyramid. Consumers that eat producers, such as squirrels, make up the next trophic level. Consumers such as hawks that eat other consumers make up the highest trophic level. Notice that less energy is available for consumers at each higher trophic level. As you read earlier, organisms use some of the energy they get from food for life processes. During life processes, some energy is changed to thermal energy and is transferred to the environment. Only about 10 percent of the energy available at one trophic level transfers on to the next trophic level.

Figure 15 An energy pyramid shows the amount of energy available at each trophic level.

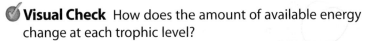 **Visual Check** How does the amount of available energy change at each trophic level?

 **Concepts in Motion** Animation

Trophic level 3
(1 percent of energy available)

Trophic level 2
(10 percent of energy available)

Trophic level 1
(100 percent of energy available)

Available energy decreases.

Lesson 3 Review

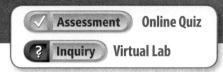
Visual Summary

Energy flows in ecosystems from producers to consumers.

Producers make their own food through the processes of photosynthesis or chemosynthesis.

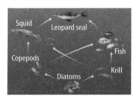
Food chains and food webs model how energy moves in ecosystems.

FOLDABLES®

Use your lesson Foldable to review the lesson. Save your Foldable for the project at the end of the chapter.

What do you think NOW?

You first read the statements below at the beginning of the chapter.

5. The Sun is the source for all energy used by living things on Earth.

6. All living things get their energy from eating other living things.

Did you change your mind about whether you agree or disagree with the statements? Rewrite any false statements to make them true.

Use Vocabulary

1 Scientists use a(n) _____ to show how energy moves in an ecosystem.

2 **Distinguish** between photosynthesis and chemosynthesis.

Understand Key Concepts

3 Which organism is a producer?
   A. cow   C. grass
   B. dog   D. human

4 **Construct** a food chain with four links.

Interpret Graphics

5 **Assess** Which trophic level has the most energy available to living things?

Ecosystem

Trophic level 3

Trophic level 2

Trophic level 1

Critical Thinking

6 **Recommend** A friend wants to show how energy moves in ecosystems. Which model would you recommend? Explain.

Math Skills
— Math Practice —

7 The plants in level 1 of a food pyramid obtain 30,000 units of energy from the Sun. How much energy is available for the organisms in level 2? Level 3?

Materials

radish seeds

sand, gravel, potting soil, and humus

paper towel

planting cups

thermometer

permanent marker

magnifying lens

250-mL beaker

metric ruler

Safety

How does soil type affect plant growth?

Plants are producers that make food through the process of photosynthesis. In order for plants to grow, they need the right balance of abiotic factors, such as light, water, air, and soil. Different plants grow well in different conditions. In this lab, you will determine which type of soil is best for growing radishes.

Ask a Question

How does the type of soil affect plant growth?

Make Observations

1. Read and complete a lab safety form.

2. Use the permanent marker to label each cup with the type of soil it will contain: *sand, gravel, potting soil, paper towel,* or *humus.*

3. Add the appropriate soil to each cup to within a thumb's width from the top. Fold the paper towel in half and place it around the inside of the cup.

4. Evenly space four seeds in each cup. Use a pencil tip to push each seed about 1 cm under the surface of the soil.

5. After planting all seeds, use the beaker to add the same amount of water to each cup. Record the amount of water used.

6. Set the cups in the same location, such as a bright windowsill. Place a thermometer next to the cups. The seeds should germinate within several days. Use a magnifying lens to observe your radish seedlings.

7. Record your observations in your Science Journal. Be sure to measure, write, draw, and label what you observe. Also, keep a record of other abiotic factors that affect the plant, such as temperature, amount of water, and hours of light.

Form a Hypothesis

8. Based on your observations, form a hypothesis regarding the effect of soil type on plant growth.

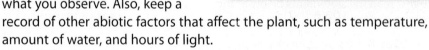

⑦ Type of soil	Observations (Day 1)	Observations (Day 4)	Observations (Day 7)
Sand			
Gravel			
Potting soil			
Humus			
Paper towel			

Test Your Hypothesis

9 Repeat your observations of the radish seedlings. Make at least four more sets of observations over the next two weeks.

10 About three weeks after the planting date, harvest your crop.

11 Gently pull up the radishes to see how they have developed underground.

12 Record data about your radish crop. How many radishes did you harvest from each cup? How big were they? What color?

Analyze and Conclude

13 **Describe** the abiotic conditions under which your plants grew. Include information about temperature, hours of sunlight, water, and soil type.

14 **Compare and Contrast** Share your data with other groups. Compare and contrast rates of growth and size of harvest.

15 **The Big Idea** Based on the class data, how does soil affect plant growth? Which type of soil is best for growing radishes?

Communicate Your Results

Combine the class data and create a graph that shows the size of plants in different soil types. Which soil produced the biggest plants? Which produced the fewest plants?

 **Inquiry Extension**

Think of another variable besides soil type that might affect the way that radish plants grow. Develop and conduct an experiment to test this variable.

Lab Tips

☑ Do not eat your radish seeds or plants.

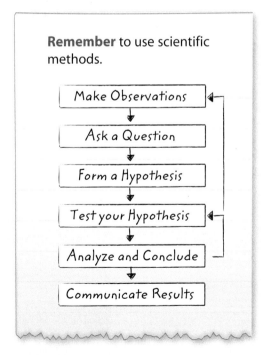

Remember to use scientific methods.

Make Observations
↓
Ask a Question
↓
Form a Hypothesis
↓
Test your Hypothesis
↓
Analyze and Conclude
↓
Communicate Results

Living things interact with and depend on each other and on the nonliving things in an ecosystem.

Key Concepts Summary 🔑	Vocabulary
Lesson 1: Abiotic Factors • The **abiotic factors** in an environment include sunlight, temperature, climate, air, water, and soil. 	**ecosystem** p. 707 **biotic factor** p. 707 **abiotic factor** p. 707 **climate** p. 708 **atmosphere** p. 709
Lesson 2: Cycles of Matter  • Matter such as oxygen, nitrogen, water, carbon, and minerals moves in cycles in the ecosystem.	**evaporation** p. 714 **condensation** p. 714 **precipitation** p. 714 **nitrogen fixation** p. 716
Lesson 3: Energy in Ecosystems • Energy flows through ecosystems from producers to consumers. • **Food chains**, **food webs**, and **energy pyramids** model the flow of energy in ecosystems. 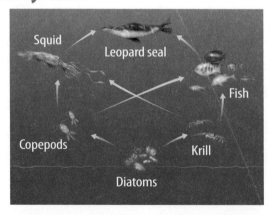	**photosynthesis** p. 724 **chemosynthesis** p. 724 **food chain** p. 726 **food web** p. 727 **energy pyramid** p. 728

FOLDABLES® Chapter Project

Assemble your lesson Foldables as shown to make a Chapter Project. Use the project to review what you have learned in this chapter.

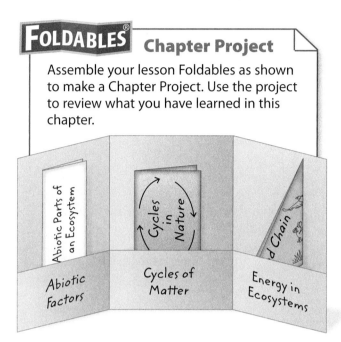

Use Vocabulary

1 Distinguish between climate and atmosphere.

2 The atmosphere is made mainly of the gases _____ and _____.

3 Living organisms in an ecosystem are called _____, while the nonliving things are called _____.

4 The process of converting nitrogen in the air into a form that can be used by living organisms is called _____ _____.

5 Use the word *precipitation* in a complete sentence.

6 Define *condensation* in your own words.

7 How does a food chain differ from a food web?

8 The process of _____ uses energy from the Sun.

9 Define *chemosynthesis* in your own words.

Link Vocabulary and Key Concepts

(((O))) **Concepts in Motion** **Interactive Concept Map**

Copy this concept map, and then use vocabulary terms from the previous page and other terms from the chapter to complete the concept map.

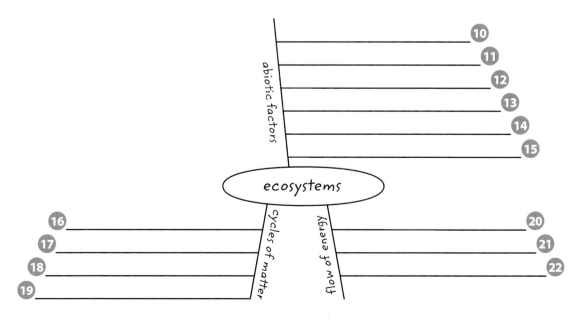

Chapter 20 Review

Understand Key Concepts 🔑

1. What is the source of most energy on Earth?
   A. air
   B. soil
   C. the Sun
   D. water

2. Which is a biotic factor in an ecosystem?
   A. a plant living near a stream
   B. the amount of rainfall
   C. the angle of the Sun
   D. the types of minerals present in soil

3. Study the energy pyramid shown here.

   Which organism might you expect to find at trophic level I?
   A. fox
   B. frog
   C. grass
   D. grasshopper

4. Which includes both an abiotic and a biotic factor?
   A. a chicken laying an egg
   B. a deer drinking from a stream
   C. a rock rolling down a hill
   D. a squirrel eating an acorn

5. Which process helps keep temperatures on Earth from becoming too hot or too cold?
   A. condensation
   B. global warming
   C. greenhouse effect
   D. nitrogen fixation

6. During the carbon cycle, _____ take in carbon dioxide from the atmosphere.
   A. animals
   B. consumers
   C. decomposers
   D. plants

7. Which is true of the amount of matter in ecosystems?
   A. It decreases over time.
   B. It increases over time.
   C. It remains constant.
   D. Scientists cannot determine how it changes.

8. Which process is occurring at the location indicated by the arrow?

   A. condensation
   B. nitrogen fixation
   C. precipitation
   D. transpiration

9. Which best represents a food chain?
   A. Sun ➔ rabbit ➔ fox ➔ grass
   B. Sun ➔ grass ➔ rabbit ➔ fox
   C. fox ➔ grass ➔ rabbit ➔ Sun
   D. grass ➔ rabbit ➔ fox ➔ Sun

10. A person who ate a salad made of lettuce, tomatoes, cheese, and ham is a(n)
    A. carnivore.
    B. detritivore.
    C. herbivore.
    D. omnivore.

Critical Thinking

11 **Compare and contrast** the oxygen cycle and the nitrogen cycle.

12 **Create** a plan for making an aquatic ecosystem in a jar. Include both abiotic and biotic factors.

13 **Recommend** a strategy for decreasing the amount of carbon dioxide in the atmosphere.

14 **Role-Play** Working in a group, perform a skit about organisms living near a hydrothermal vent. Be sure to include information about how the organisms obtain energy.

15 **Assess** the usefulness of models as tools for studying ecosystems.

16 Study the food web below. **Classify** each organism according to what it eats.

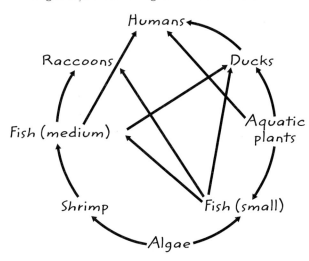

17 **Predict** what would happen if all the nitrogen-fixing bacteria in an ecosystem were removed.

Writing in Science

18 **Write** an argument for or against the following statement. *The energy humans use in cars originally came from the Sun.*

REVIEW THE BIG IDEA

19 Describe an interaction between a living thing and a nonliving thing in an ecosystem.

20 How might the ram interact with nonliving things in its environment?

Math Skills ×÷

Review
— Math Practice —

Use Percentages

21 A group of plankton, algae, and other ocean plants absorb 150,000 units of energy.

  a. How much energy is available for the third trophic level?

  b. How much energy would remain for a fourth trophic level?

22 Some organisms, such as humans, are omnivores. They eat both producers and consumers. How much more energy would an omnivore get from eating the same mass of food at the first trophic level than at the second trophic level?

Standardized Test Practice

Record your answers on the answer sheet provided by your teacher or on a sheet of paper.

Multiple Choice

1 In which process do producers use chemical energy and make food?

  A chemosynthesis

  B fermentation

  C glycolysis

  D photosynthesis

Use the image below to answer question 2.

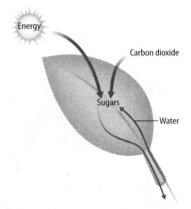

2 What process is shown above?

  A chemosynthesis

  B decomposition

  C nitrogen fixation

  D photosynthesis

3 What organisms help break down dead leaves in an ecosystem?

  A carnivores

  B detritivores

  C herbivores

  D omnivores

4 Which process converts atmospheric nitrogen to a form organisms can use?

  A absorption

  B fixation

  C retention

  D stabilization

Use the diagram below to answer question 5.

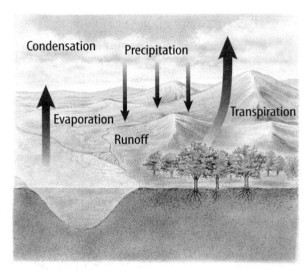

5 Which step of the water cycle shown above occurs in plants?

  A condensation

  B evaporation

  C precipitation

  D transpiration

6 Which is true of energy in ecosystems?

  A It never changes form.

  B It is both created and destroyed.

  C It flows in one direction.

  D It follows a cycle pattern.

7 Which organism would most likely appear at the top of an energy pyramid?

  A grass

  B hawk

  C mouse

  D snake

Use the diagram below to answer questions 8 and 9.

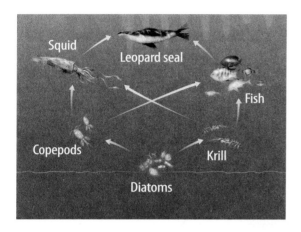

8 How does energy move in the food web pictured above?

  A from leopard seal to squid

  B from diatoms to krill

  C from fish to krill

  D from squid to diatoms

9 Which is an example of a food chain shown above?

  A diatoms → krill → leopard seal

  B fish → krill → squid

  C diatoms → krill → fish

  D squid → fish → leopard seal

10 During which process is oxygen gas released into the atmosphere?

  A chemosynthesis

  B decomposition

  C photosynthesis

  D transpiration

Constructed Response

11 Most ecosystems contain six nonliving factors: atmosphere, climate, soil, temperature, sunlight, and water. Briefly explain how each factor affects the life of a large predator, such as a jaguar, in a jungle ecosystem.

Use the diagram below to answer questions 12 and 13.

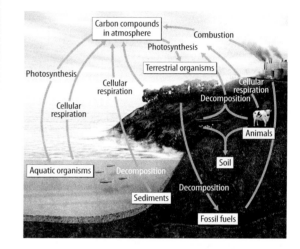

12 Using the image of the carbon cycle shown above, identify three locations of carbon other than the atmosphere. What form does the carbon take in each example?

13 Describe a biological process of the carbon cycle that removes carbon from the atmosphere and a biological process that adds carbon dioxide to the atmosphere.

NEED EXTRA HELP?													
If You Missed Question...	1	2	3	4	5	6	7	8	9	10	11	12	13
Go to Lesson...	3	2	3	2	2	3	3	3	3	2	1	2	2

Populations and Communities

THE BIG IDEA
How do populations and communities interact and change?

Inquiry Too Many Pigeons?

This group of pigeons does not depend only on the environment for food. Tourists visiting the area also feed the pigeons. Because so much food is available, more pigeons than normal live in this part of the city.

- Do you think this large number of pigeons affects other organisms in the area?

- How do you think groups of pigeons and other organisms interact and change?

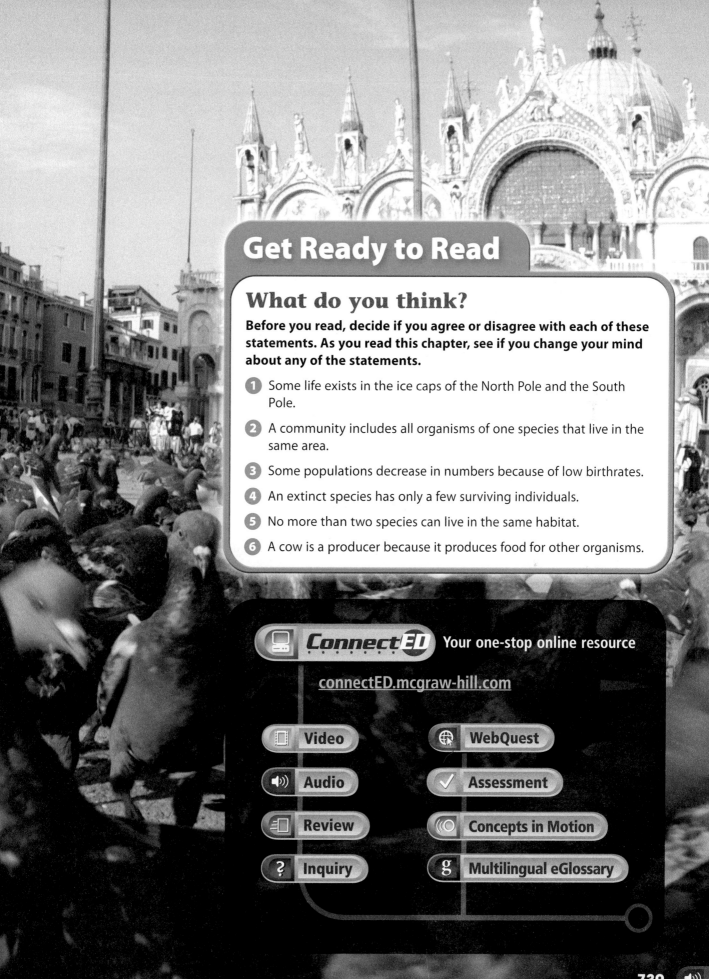

Get Ready to Read

What do you think?

Before you read, decide if you agree or disagree with each of these statements. As you read this chapter, see if you change your mind about any of the statements.

1 Some life exists in the ice caps of the North Pole and the South Pole.

2 A community includes all organisms of one species that live in the same area.

3 Some populations decrease in numbers because of low birthrates.

4 An extinct species has only a few surviving individuals.

5 No more than two species can live in the same habitat.

6 A cow is a producer because it produces food for other organisms.

ConnectED Your one-stop online resource

connectED.mcgraw-hill.com

Video

WebQuest

Audio

Assessment

Review

Concepts in Motion

Inquiry

Multilingual eGlossary

Lesson 1

Reading Guide

Key Concepts 🔑

ESSENTIAL QUESTIONS

- What defines a population?
- What factors affect the size of a population?

Vocabulary

biosphere p. 741

community p. 742

population p. 742

competition p. 743

limiting factor p. 743

population density p. 744

biotic potential p. 744

carrying capacity p. 745

g Multilingual eGlossary

Populations

Inquiry Looking for Something?

Meerkats live in family groups. They help protect each other by watching for danger from eagles, lions, and other hunters of the Kalahari Desert. What other ways might the meerkats interact?

How many times do you interact?

Every day, you interact with other people in different ways, including talking, writing, or shaking hands. Some interactions involve just one other person, and others happen between many people. Like humans, other organisms interact with each other in their environment.

1. Make a list in your Science Journal of all the ways you have interacted with other people today.

2. Use a **highlighter** to mark the interactions that occurred between you and one other person.

3. Use a **highlighter** of another color to mark interactions that occurred among three or more people.

Think About This

1. Were your interactions mainly with one person or with three or more people?

2. 🔑 **Key Concept** How might your interactions change if the group of people were bigger?

The Biosphere and Ecological Systems

Imagine flying halfway around the world to Africa. When your plane flies over Africa, you might see mountains, rivers, grasslands, and forests. As you get closer to land, you might see a herd of elephants at a watering hole. You also might see a group of meerkats, like the ones on the previous page.

Now imagine hiking through an African forest. You might see monkeys, frogs, insects, spiders, and flowers. Maybe you catch sight of crocodiles sunning themselves by a river or birds perching on trees.

You are exploring Earth's **biosphere** (BI uh sfir)—*the parts of Earth and the surrounding atmosphere where there is life.* The biosphere includes all the land of the continents and islands. It also includes all of Earth's oceans, lakes, and streams, as well as the ice caps at the North Pole and the South Pole.

Parts of the biosphere with large amounts of plants or algae often contain many other organisms as well. The biosphere's distribution of chlorophyll, a green pigment in plants and algae, is shown in **Figure 1.**

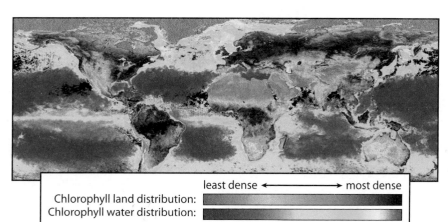

Chlorophyll land distribution:
Chlorophyll water distribution:

least dense ⟷ most dense

Figure 1 The colors in this satellite image represent the densities of chlorophyll, a green pigment found in plants and algae.

✓ **Visual Check** Why might the North Pole have very little green?

What is a population?

The Kalahari Desert in Africa is a part of the Earth's biosphere. A wildlife refuge in the Kalahari Desert is home to several groups of meerkats. Meerkats are small mammals that live in family groups and help each other care for their young.

Meerkats rely on interactions among themselves to survive. They sleep in underground burrows at night and hunt for food during the day. They take turns standing upright to watch for danger and call out warnings to others.

Meerkats are part of an ecosystem, as shown in **Figure 2**. An ecosystem is a group of organisms that lives in an area at one time, as well as the climate, soil, water, and other nonliving parts of the environment. The Kalahari Desert is an ecosystem. The study of all ecosystems on Earth is ecology.

Many species besides meerkats live in the Kalahari Desert. They include scorpions, spiders, insects, snakes, and birds such as eagles and owls. Also, large animals like zebras, giraffes, and lions live there. Plants that grow in the Kalahari Desert include shrubs, grasses, small trees, and melon vines. Together, all these plants, animals, and other organisms form a community. *A **community** is all the populations of different species that live together in the same area at the same time.*

All the meerkats in this refuge form a population. *A **population** is all the organisms of the same species that live in the same area at the same time.* A species is a group of organisms that have similar traits and are able to produce fertile offspring.

 Key Concept Check What defines a population?

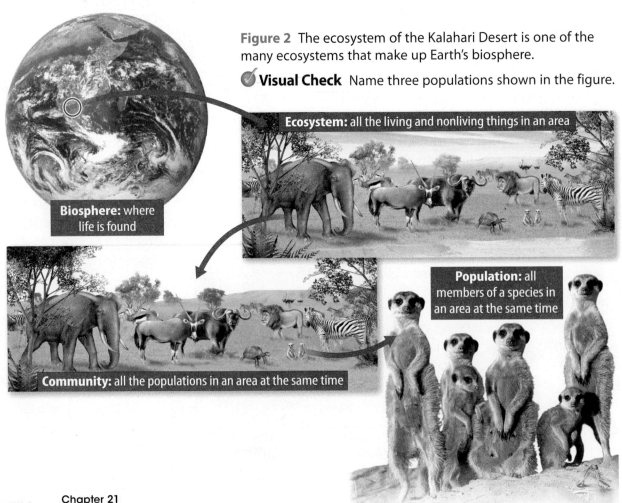

Figure 2 The ecosystem of the Kalahari Desert is one of the many ecosystems that make up Earth's biosphere.

Visual Check Name three populations shown in the figure.

Biosphere: where life is found

Ecosystem: all the living and nonliving things in an area

Community: all the populations in an area at the same time

Population: all members of a species in an area at the same time

Competition

At times, not enough food is available for every organism in a community. Members of a population, including those in the Kalahari Desert, must compete with other populations and each other for enough food to survive. **Competition** *is the demand for resources, such as food, water, and shelter, in short supply in a community.* When there are not enough resources available to survive, there is more competition in a community.

Population Sizes

If the amount of available food decreases, what do you think happens to a population of meerkats? Some meerkats might move away to find food elsewhere. Female meerkats cannot raise as many young. The population becomes smaller. If there is plenty of food, however, the size of the population grows larger as more meerkats survive to adulthood and live longer. Changes in environmental factors can result in population size changes.

Limiting Factors

Environmental factors, such as available water, food, shelter, sunlight, and temperature, are possible limiting factors for a population. *A* **limiting factor** *is anything that restricts the size of a population.* Available sunlight is a limiting factor for most organisms. If there is not enough sunlight, green plants cannot make food by photosynthesis. Organisms that eat plants are affected if little food is available.

Temperature is a limiting factor for some organisms. When the temperature drops below freezing, many organisms die because it is too cold to carry out their life functions. Disease, predators—animals that eat other animals—and natural disasters such as fires or floods are limiting factors as well.

 Key Concept Check What factors affect the size of a population?

inquiry MiniLab 15 minutes

What are limiting factors?

Certain factors, called limiting factors, can affect the size of a population.

1. Read and complete a lab safety form.
2. Your teacher will divide your class into groups.
3. Using a **meterstick** and **masking tape,** mark a 1-m square on the floor. Place a piece of paper in the middle of the square.
4. All members of your group will stand entirely within the square. While one member keeps time with a **stopwatch,** members of the group will write the alphabet on the sheet of paper one at a time.

5. In your Science Journal, calculate the average time it took each person to write the alphabet.

Analyze and Conclude

1. **Describe** how the space limitations affected each member's ability to complete the task.

2. 🔑 **Key Concept** What functions must an organism perform that can be limited by the amount of available space?

Figure 3 A sedated lynx is fitted with a radio collar and then returned to the wild.

WORD ORIGIN · · · · · · · · · · · ·

population
from Latin *populus*, means "inhabitants"

density
from Latin *densus*, means "thick, crowded"

FOLDABLES

Make a horizontal half book and label it as shown. Use it to organize your notes on the relationship between population size and carrying capacity in an ecosystem.

Carrying Capacity

Measuring Population Size

Sometimes it is difficult to determine the size of a population. How would you count scampering meerkats or wild lynx? One method used to count and monitor animal populations is the capture-mark-and-release method. The lynx in **Figure 3** is a member of a population in Poland that is monitored using this method. Biologists using this method sedate animals and fit them with radio collars before releasing them back into the wild. By counting how many observed lynx are wearing collars, scientists can estimate the size of the lynx population. Biologists also use the collars to track the lynx's movements and monitor their activities.

Suppose you want to know how closely together Cumberland azaleas (uh ZAYL yuhz), a type of flower, grow in the Great Smoky Mountains National Park. **Population density** *is the size of a population compared to the amount of space available.* One way of estimating population density is by sample count. Rather than counting every azalea shrub, you count only those in a representative area, such as 1 km^2. By multiplying the number of square kilometers in the park by the number of azaleas in 1 km^2, you find the estimated population density of azalea shrubs in the entire park.

 Reading Check Describe two ways you can estimate population size.

Biotic Potential

Imagine that a population of raccoons has plenty of food, water, and den space. In addition, there is no disease or danger from other animals. The only limit to the size of this population is the number of offspring the raccoons can produce. **Biotic potential** *is the potential growth of a population if it could grow in perfect conditions with no limiting factors.* No population on Earth ever reaches its biotic potential because no ecosystem has an unlimited supply of natural resources.

Carrying Capacity

What would happen if a population of meerkats reached its biotic potential? It would stop growing when it reached the limit of available resources that the ecosystem could provide, such as food, water, or shelter. *The largest number of individuals of one species that an environment can support is the* **carrying capacity.** A population grows until it reaches the carrying capacity of an environment, as shown in **Figure 4.** Disease, space, predators, and food are some of the factors that limit the carrying capacity of an ecosystem. However, the carrying capacity of an environment is not constant. It increases and decreases as the amount of available resources increases and decreases. At times, a population can temporarily exceed the carrying capacity of an environment.

 Reading Check What is carrying capacity?

Overpopulation

When the size of a population becomes larger than the carrying capacity of its ecosystem, overpopulation occurs. Overpopulation can cause problems for organisms. For example, meerkats eat spiders. An overpopulation of meerkats causes the size of the spider population in that community to decrease. Populations of birds and other animals that eat spiders also decrease when the number of spiders decreases.

Elephants in Africa's wild game parks is another example of overpopulation. Elephants searching for food caused the tree damage shown in **Figure 5.** They push over trees to feed on the uppermost leaves. Other species of animals that use the same trees for food and shelter must compete with the elephants. The loss of trees and plants can also damage soil. Trees and plants might not grow in that area again for a long time.

Reading Check How can overpopulation affect a community?

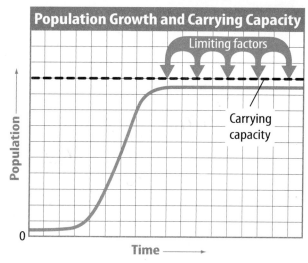

▲ **Figure 4** Carrying capacity is determined in part by limiting factors.

Visual Check What factors affect population size in the graph above?

▲ **Figure 5** An overpopulation of elephants can cause damage to trees and other plants as the herd searches for food in the community.

Lesson 1 Review

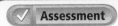

Visual Summary

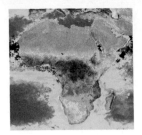

The population density of organisms, including green plants and algae, varies throughout the world.

A community is all the populations of different species that live together in the same area at the same time.

The number of individuals in a population varies as the amount of available resources varies.

FOLDABLES®

Use your lesson Foldable to review the lesson. Save your Foldable for the project at the end of the chapter.

What do you think NOW?

You first read the statements below at the beginning of the chapter.

1. Some life exists in the ice caps of the North Pole and the South Pole.

2. A community includes all organisms of one species that live in the same area.

Did you change your mind about whether you agree or disagree with the statements? Rewrite any false statements to make them true.

Use Vocabulary

1 **Define** *population.*

2 **Distinguish** between carrying capacity and biotic potential.

3 Food, water, living space, and disease are examples of _____.

Understand Key Concepts

4 **Explain** how competition could limit the size of a bird population.

5 One example of competition among members of a meerkat population is
   A. fighting over mates.
   B. warning others of danger.
   C. huddling together to stay warm.
   D. teaching young to search for food.

Interpret Graphics

6 **Sequence** Draw a graphic organizer like the one below to show the sequence of steps in one type of population study.

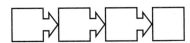

7 **Explain** the changes in population size at each point marked on the graph below.

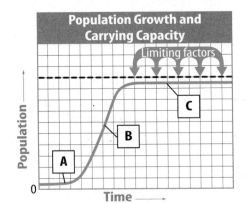

Critical Thinking

8 **Explain** Is the problem of elephants destroying trees in southern Africa overpopulation, competition, or both?

Familiar Birds
in an Unlikely Place

Howling winds blow across the Altiplano—a cold plateau high in the Andes mountain range of South America. There, you might expect to see animals such as llamas, but Felicity Arengo travels to the Altiplano to observe flamingos.

Flamingos usually are associated with tropical regions. However, three species of flamingo—James, Andean, and Chilean—are adapted to the cold, barren Altiplano. Although this region differs from tropical areas where other flamingos live, both have food sources for the birds. The plateau is dotted with salty lakes containing brine shrimp, tiny organisms that flamingos eat.

Scientists have many questions about the flamingos that visit these salty lakes. How large are the flamingo populations? How do they survive when the lakes evaporate? To answer these questions, Arengo and her team visit the lakes and count flamingos there. They also tag flamingos with radio transmitters to track their movements. They have learned that one species, the Andean flamingo, is the rarest flamingo species in the world. Additionally, when the plateau lakes freeze, many flamingos fly to lakes in the lowlands of Argentina, Bolivia, Chile, and Peru.

As human activity changes the Altiplano, flamingos that live there might be in danger. On the plateau, mining operations use and pollute lake water. In the lowlands, ranchers often drain lakes for more land to grow crops or to feed animals.

Arengo and other scientists are working with organizations to protect flamingos' habitats. Scientists have trained park rangers to monitor flamingos' reproductive activities and protect nesting colonies. As scientists collect more data and find new ways to protect flamingos' habitats, a brighter future might be in store for the flamingos of the Altiplano.

▲ **Dr. Arengo tags a flamingo with a radio transmitter. Once she releases the flamingo, satellites will track the flamingo's movement.**

▲ **Flamingos' habitats cover four countries—Argentina, Bolivia, Chile, and Peru. As a well-known species, flamingos help motivate conservation efforts in these countries.**

It's Your Turn

BRAINSTORM With classmates, choose an ecosystem in your area that is in need of conservation. Brainstorm what animal would make a good species to represent the ecosystem, and create a poster designed to raise awareness.

Changing Populations

Reading Guide

Key Concepts
ESSENTIAL QUESTIONS

- How do populations change?

- Why do human populations change?

Vocabulary

birthrate p. 749

death rate p. 749

extinct species p. 751

endangered species p. 751

threatened species p. 751

migration p. 752

g Multilingual eGlossary

Video BrainPOP®

Inquiry Same Mother?

Have you ever seen newly hatched baby spiders? Baby spiders can have hundreds or even thousands of brothers and sisters. What keeps the spider population from growing out of control?

What events can change a population?

Populations can be affected by human-made and environmental changes, such as floods or a good growing season. A population's size can increase or decrease in response to these changes.

1. Read and complete a lab safety form.

2. Record in your Science Journal the number of **counting objects** you have been given. Each object represents an organism, and all the objects together represent a population.

3. Turn over one of the **event cards** you were given and follow the instructions on the card. Determine the event's impact on your population.

4. Repeat step 3 for four more "seasons," or turns.

Think About This

1. Compare the size of your population with other groups. Do you all have the same number of organisms at the end of five seasons?

2. 🔑 **Key Concept** What effect did the different events have on your population?

How Populations Change

Have you ever seen a cluster of spider eggs? Some female spiders lay hundreds or even thousands of eggs in their lifetime. What happens to a population of spiders when a large group of eggs hatches all at once? The population suddenly becomes larger. It doesn't stay that way for long, though. Many spiders die or become food, like the one being eaten in **Figure 6,** before they grow enough to reproduce. The size of the spider population increases when the eggs hatch but decreases as the spiders die.

A population change can be measured by the population's birthrate and death rate. *A population's* **birthrate** *is the number of offspring produced over a given time period. The* **death rate** *is the number of individuals that die over the same time period.* If the birthrate is higher than the death rate, the population increases. If the death rate is higher than the birthrate, the population decreases.

Figure 6 Spiders have a high birthrate, but they usually have a high death rate too. Many spiders die or are eaten before they can reproduce.

Exponential Growth

SCIENCE USE v. COMMON USE

exponential

Science Use a mathematical expression that contains a constant raised to a power, such as 2^3 or x^2

Common Use in great amounts

When a population is in ideal conditions with unlimited resources, it grows in a pattern called exponential growth. During exponential growth, the larger a population gets, the faster it grows. *E. coli* bacteria are microscopic organisms that undergo exponential growth. This population doubles in size every half hour, as shown in **Figure 7**. It takes only 10 hours for the *E. coli* population to grow from one organism to more than 1 million. Exponential growth cannot continue for long. Eventually, limiting factors stop population growth.

Exponential Population Growth

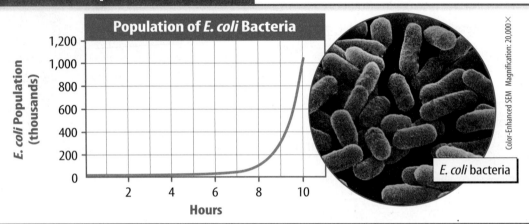

Population of *E. coli* Bacteria

E. coli Population (thousands)

Hours

Color-Enhanced SEM Magnification: 20,000×

E. coli bacteria

Figure 7 When grown in the laboratory, this population of *E. coli* bacteria is given everything it needs to briefly achieve exponential growth.

WORD ORIGIN

extinct

from Latin *extinctus*, means "extinguish"

Population Size Decrease

Population size can increase, but it also can decrease. For example, a population of field mice might decrease in size in the winter because there is less food. Some mice might not be able to find enough food and will starve. More mice will die than will be born, so the population size decreases. When food is plentiful, the population size usually increases.

Natural disasters such as floods, fires, or volcanic eruptions also affect population size. For example, if a hurricane rips away part of a coral reef, the populations of coral and other organisms that live on the reef also decrease in size.

Disease is another cause of population decrease. In the mid-1900s, Dutch elm disease spread throughout the United States and destroyed many thousands of elm trees. Because of the disease, the size of population of elm trees decreased.

Predation—the hunting of organisms for food—also reduces population size. For example, a farmer might bring cats into a barn to reduce the size of a mouse population.

✓ **Reading Check** What are four reasons that a population might decrease in size?

Extinction If populations continue to decrease in numbers, they disappear. *An extinct species is a species that has died out and no individuals are left.* Extinctions can be caused by predation, natural disasters, or damage to the environment.

Some extinctions in Earth's history were large events that involved many species. Most scientists think the extinction of the dinosaurs about 65 million years ago was caused by a meteorite crashing into Earth. The impact would have sent tons of dust into the atmosphere, blocking sunlight. Without sunlight, plants could not grow. Animals, such as dinosaurs, that ate plants probably starved.

Most extinctions involve fewer species. For example, New Zealand was once home to a large, flightless bird called the giant moa, as shown in **Figure 8.** Humans first settled these islands about 700 years ago. They hunted the moa for food. As the size of the human population increased, the size of the moa population decreased. Within 200 years, all the giant moas had been killed and the species became extinct.

Endangered Species The mountain gorillas shown in **Figure 8** are an example of a species that is endangered. *An* **endangered species** *is a species whose population is at risk of extinction.*

Threatened Species California sea otters almost became extinct in the early 1900s due to overhunting. In 1977, California sea otters were classified as a **threatened species**—*a species at risk, but not yet endangered.* Laws were passed to protect the otters and by 2007 there were about 3,000 sea otters. Worldwide, there are more than 4,000 species that are classified as endangered or threatened.

Reading Check What is the difference between an endangered species and a threatened species?

Figure 8 Organisms are classified as extinct, endangered, or threatened.

Extinct The giant moa, a large bird that was nearly four meters tall, was hunted to extinction.

Endangered Just over 700 mountain gorillas remain in the wild in Africa.

Threatened California sea otters are at risk of becoming endangered because there are so few of them remaining in the wild.

How does migration affect population size?

Your class will model a population of birds that migrates during the fall and the spring.

1 Read and complete a lab safety form.

2 Begin at the summer station. Record in your Science Journal the size of the bird population represented by your class. When your teacher signals, move to the fall station.

3 Pick a piece of **paper** out of the **jar.** If the paper has a minus sign, drop out of the game. If it has a plus sign, bring a classmate into the game.

4 Record the size of the remaining population at the fall station.

5 Migrate to the winter station and repeat steps 3 and 4. Move two more times and repeat for spring and summer.

Analyze and Conclude

1. **Draw Conclusions** What might happen if the birds did not migrate each year?

2. 🔑 **Key Concept** How did the population change throughout the year?

Migration

Canada

United States

ATLANTIC OCEAN

Gulf of Mexico

Bahamas

Puerto Rico

Cuba

Jamaica   Haiti   Dominican Republic

Caribbean Sea

Figure 9 During the winter, humpback whales mate and give birth in warm ocean waters near the Bahamas. In the summer, they migrate north to food-rich waters along the coast of New England.

Movement

Populations also change when organisms move from place to place. When an animal population becomes overcrowded, some individuals might move to find more food or living space. For example, zebras might overgraze an area and move to areas that are not so heavily grazed.

Plant populations can also move from place to place. Have you ever blown on a dandelion puff full of seeds? Each tiny dandelion seed has a feathery part that enables it to be carried by the wind. Wind often carries seeds far from their parent plants. Animals also help spread plant seeds. For example, some squirrels and woodpeckers collect acorns. They carry the acorns away and store them for a future food source. The animal forgets some acorns, and they sprout and grow into new trees far from their parent trees.

Migration Sometimes an entire population moves from one place to another and later returns to its original location. **Migration** *is the instinctive seasonal movement of a population of organisms from one place to another.* Ducks, geese, and monarch butterflies are examples of organisms that migrate annually. Some fish, frogs, insects, and mammals—including the whales described in **Figure 9**—migrate to find food and shelter.

🔑 **Key Concept Check** List three ways populations change.

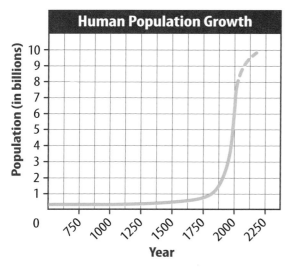

Human Population Growth

◀ **Figure 10** The human population has grown faster in the past 150 years than at any time in Earth's history.

✓ **Visual Check** How does this population curve compare with the graph of the *E. coli* population in **Figure 7?**

Human Population Changes

Human population size is affected by the same three factors that determine the sizes of all populations—birthrate, death rate, and movement. But, unlike other species, humans have developed ways to increase the carrying capacity of their environment. Improved crop yields, domesticated farm animals, and timely methods of transporting foods and other resources enable people to survive in all types of environments.

Scientists **estimate** that there were about 300 million humans on Earth a thousand years ago. Today there are more than 6 billion humans on Earth, as shown in **Figure 10.** By 2050 there could be over 9 billion. No one knows when the human population will reach Earth's carrying capacity. However, some scientists estimate Earth's carrying capacity is about 11 billion.

As the human population grows, people need to build more houses and roads and clear more land for crops. This means less living space, food, and other resources for other species. In addition, people use more energy to heat and cool homes; to fuel cars, airplanes, and other forms of transportation; and to produce electricity. This energy use contributes to pollution that affects other populations.

One example of the consequences of human population growth is the destruction of tropical forests. Each year, humans clear thousands of acres of tropical forest to make room for crops and livestock, as shown in **Figure 11.** Clearing tropical forests is harmful because these forests contain a large variety of species that are not in other ecosystems.

✓ **Reading Check** Explain how human population growth affects other species.

ACADEMIC VOCABULARY

estimate

(verb) to determine roughly the size, nature, or extent of something

Figure 11 Tropical forests are cleared for crops and livestock. The habitats of many organisms are destroyed, resulting in many species becoming endangered or extinct. ▼

Figure 12 Before vaccinations, many children died in infancy. The use of vaccines has significantly reduced death rates.

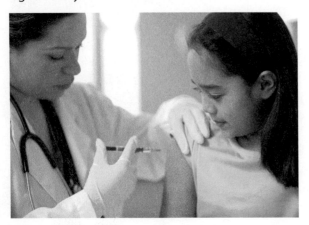

Population Size Increase

Do you know anyone who is more than 100 years old? In 2006, almost 80,000 people living in the United States were at least 100 years old. People are living longer today than in previous generations, and more children reach adulthood. Recall that when the birthrate of a population is higher than its death rate, the population grows. There are several factors that keep the human birthrate higher than its death rate. Some of these factors are discussed below.

Food For some, finding food might be as easy as making a trip to the grocery store, but not everyone can get food as easily. Advances in agriculture have made it possible to produce food for billions of people.

Resources Fossil fuels, cloth, metals, foods, and many other materials are easily transported around the world by planes, trains, trucks, or boats. Today, people have access to more resources because of better transportation methods.

Sanitation As recently as 100 years ago, diseases such as typhoid, cholera, and diphtheria were major causes of death. These diseases spread through unclean water supplies and untreated sewage. Modern water treatment technologies have reduced the occurrence of many diseases. Less expensive and more effective cleaning products are now available to help prevent the spread of disease-causing organisms. As a result, deaths from these illnesses are less common in many countries.

Medical Care Modern medical care is keeping people alive and healthy longer than ever before. As shown in **Figure 12,** scientists have developed vaccines, antibiotics, and other medicines that prevent and treat disease. As a result, fewer people get sick, and human death rates have decreased. Medical technologies and new medicines help people survive heart attacks, cancer, and other major illnesses.

Decreases in Human Population Size

Human populations in some parts of the world are decreasing in size. Diseases such as AIDS and malaria cause high death rates in some countries. Severe drought has resulted in major crop failures and lack of food. Floods, earthquakes, and other natural disasters can cause the deaths of hundreds or even thousands of people at a time. Damage from disasters, such as the damage shown in **Figure 13,** can keep people from living in the area for a long time. All of these factors cause decreases in human population sizes in some areas.

 Reading Check What are three events that can decrease human population size?

Population Movement

Have you ever moved to a different city, state, or country? The size of a human population changes as people move from place to place. The graph in **Figure 14** shows the percentages of each kind of move people make. Like other organisms, populations of humans might move when more resources become available in a different place.

Did your parents, grandparents, or great grandparents come to the United States from another country? Immigration takes place when organisms move into an area. Most of the U.S. population is descended from people who immigrated from Europe, Africa, Asia, and Central and South America.

 Key Concept Check What makes human populations increase or decrease in size?

▲ **Figure 13** Natural disasters such as a tsunami can cause severe damage to people's homes, as well as drastically reduce the population size.

FOLDABLES

Make a horizontal two-tab book and label it as shown. Use it to summarize why human populations change in size.

| Human Population Increase | Human Population Decrease |

Types of Moves, 2004–2005

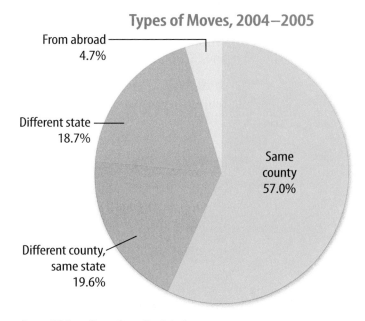

From abroad
4.7%

Different state
18.7%

Same county
57.0%

Different county, same state
19.6%

◀ **Figure 14** Populations can move between counties and states or even from another country.

 Visual Check Which type of move did the largest percentage of the population make?

Source: U.S. Census Bureau, Current Population Survey, 2005 Annual Social and Economic Supplement.

Lesson 2 Review

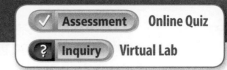

Visual Summary

The birthrate and the death rate of any population affects its population size.

The giant moa is classified as an extinct species because there are no surviving members.

A population that is at risk but not yet endangered is a threatened species.

FOLDABLES®

Use your lesson Foldable to review the lesson. Save your Foldable for the project at the end of the chapter.

What do you think NOW?

You first read the statements below at the beginning of the chapter.

3. Some populations decrease in numbers because of low birthrates.

4. An extinct species has only a few surviving individuals.

Did you change your mind about whether you agree or disagree with the statements? Rewrite any false statements to make them true.

Use Vocabulary

1. **Define** *endangered species* in your own words.

2. **Distinguish** between birthrate and death rate.

3. The instinctive movement of a population from one place to another is _____.

Understand Key Concepts

4. Rabbits move into a new field where there is plenty of room to dig new burrows. This is an example of
   - **A.** overpopulation.
   - **B.** immigration.
   - **C.** carrying capacity.
   - **D.** competition.

Interpret Graphics

5. **Summarize** Copy and fill in the graphic organizer below to identify the three major factors that affect population size.

Critical Thinking

6. **Predict** what could happen to the size of the human population if a cure for all cancers were discovered.

7. **Recommend** an action humans could take to help prevent the extinction of tropical organisms.

Math Skills

Review — Math Practice

Use the graph in the Skill Practice on the next page to answer these questions.

8. What does each unit on the *y*-axis represent?

9. What does the lowest point on the blue line represent?

How do populations change in size?

Birthrate and death rate change the size of a population. In the 1700s the death rate of sea otters in central California was extremely high because many people hunted them. By the 1930s only about 50 sea otters remained. Today, the Marine Mammal Protection Act protects sea otters from being hunted. Every spring, scientists survey the central California Coast to determine the numbers of adult and young sea otters (called pups) in the population. The numbers on the graph indicate population sizes at the end of a breeding season.

Learn It

Most scientists collect some type of data when testing a hypothesis. Once data are collected, scientists look for patterns or trends in the data and draw conclusions. This process is called **interpreting data.**

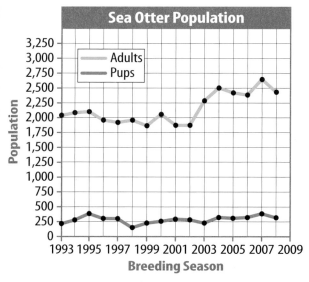

Sea Otter Population

Try It

1 The above graph shows changes in adult and pup sea otter populations over many years. Assume that the number of pups seen during the survey represents all the pups that were born and survived in one year—the birthrate. For example, in the 1997 breeding season, the birthrate was 300.

2 In your Science Journal, make a table showing the population size and the birthrate for the 2001 breeding season. Repeat for 2002, 2003, and 2004.

3 In each breeding season, the population increases by the number of pups born and decreases by the number of sea otters that die. Use the following equation to find the death rate for 2002.

Death rate in 2002 = population size in 2001 + birthrate in 2002 − population size in 2002

Apply It

4 Calculate the death rate in 2004 and compare it to the death rate in 2002.

5 What environmental factors might account for the difference in the death rate between 2002 and 2004?

6 How do you think the population size will change in 2009 and 2010?

7 🔑 **Key Concept** Determine how the birthrate compared to the death rate in 2002 and 2004. Explain how these rates affected the population sizes in 2002 and 2004.

Lesson 3

Communities

Reading Guide

Key Concepts
ESSENTIAL QUESTIONS

- What defines a community?
- How do the populations in a community interact?

Vocabulary

habitat p. 759

niche p. 759

producer p. 760

consumer p. 760

symbiosis p. 763

mutualism p. 763

commensalism p. 764

parasitism p. 764

g Multilingual eGlossary

Inquiry Time for Lunch?

This Hoopoe (HOO poo) has captured its next meal. Some of the energy needed by this bird for its life processes will come from the energy stored in the body of the lizard. Where did the lizard get its energy?

What are the roles in your school community?

Within a community, different organisms have different roles. Trees produce their own food from the environment. Then, they become food for other organisms. Mushrooms break down dead organisms and make the nutrients useful to other living things. Think about the members of your school community such as the students, teachers, and custodians. What roles do they have?

1. Draw a table with two columns in your Science Journal. Label one column *Community Member* and the other column *Role in the Community*.

2. Fill in the table with examples from your school.

Community Member	Role in the Community
Principal	Manages school staff including students and teachers

Think About This

1. Are there any community members who have more than one role?

2. What is your role in the school community?

3. **Key Concept** Explain how it is beneficial for members of a community to have different roles.

Communities, Habitats, and Niches

High in a rain forest tree, a two-toed sloth munches leaves. Ants crawl on a branch, carrying away a dead beetle. Two birds build a nest. A flowering vine twists around the tree trunk. These organisms are part of a rain forest community. You read in Lesson 1 that a community is made up of all the species that live in the same ecosystem at the same time.

The place within an ecosystem where an organism lives is its **habitat.** A habitat, like the one in **Figure 15,** provides all the resources an organism needs, including food and shelter. A habitat also has the right temperature, water, and other conditions the organism needs to survive.

The rain forest tree described above is a habitat for sloths, insects, birds, vines, and many other species. Each species uses the habitat in a different way. *A* **niche** (NICH) *is what a species does in its habitat to survive.* For example, butterflies feed on flower nectar. Sloths eat leaves. Ants eat insects or plants. These species have different niches in the same environment. Each organism shown in **Figure 15** has its own niche on the tree. The plants anchor themselves to the tree and can capture more sunlight. Termites use the tree for food.

Key Concept Check What is a community?

WORD ORIGIN · · · · · · · · · · ·

habitat
from Latin *habitus*, means "to live, dwell"

Figure 15 This tree trunk is a habitat for ferns.

How can you model a food web?

Populations interact through feeding relationships. A food web shows overlapping feeding relationships in a community.

1. Read and complete a lab safety form.
2. On a sheet of **paper,** make a list of at least 10 different organisms within a community of your choice. Include a variety of producers and consumers.
3. Use **scissors** to cut out the name of each organism on your list.
4. **Glue** the names onto a piece of **construction paper.**

5. Use **yarn** and glue to connect organisms that have feeding relationships. For example, a piece of yarn would connect a rabbit and grass.

Analyze and Conclude

1. **Use Models** Add the label *Sun* to your model. Which organisms would be connected to the Sun?
2. **Infer** Imagine that you removed three organisms from your food web. How would this affect the community?
3. **Key Concept** Which organisms in your model interact through feeding relationships?

Energy in Communities

Sloths are the slowest mammals on Earth. They hardly make a sound, and they sleep 15 to 20 hours a day. Squirrel monkeys, however, chatter as they swing through treetops hunting for fruit, insects, and eggs. Sloths might appear to use no energy at all. However, sloths, squirrel monkeys, and all other organisms need energy to live. All living things use energy and carry out life processes such as growth and reproduction.

Energy Roles

How an organism obtains energy is an important part of its niche. Almost all the energy available to life on Earth originally came from the Sun. However, some organisms, such as those that live near deep-sea vents, are exceptions. They obtain energy from chemicals such as hydrogen sulfide.

Producers *are organisms that get energy from the environment, such as sunlight, and make their own food.* For example, most plants are producers that get their energy from sunlight. They use the process of photosynthesis and make sugar molecules that they use for food. Producers near deep-sea vents use hydrogen sulfide and carbon dioxide and make sugar molecules.

Consumers *are organisms that get energy by eating other organisms.* Consumers are also classified by the type of organisms they eat. Herbivores get their energy by eating plants. Cows and sheep are herbivores. Carnivores get their energy by eating other consumers. Harpy eagles, lions, and wolves are carnivores. Omnivores, such as most humans, get their energy by eating producers and consumers. Detritivores (dee TRI tuh vorz) get their energy by eating dead organisms or parts of dead organisms. Some bacteria and some fungi are detritivores.

Reading Check Identify a producer, an herbivore, a carnivore, and an omnivore.

Energy Flow

A food chain is a way of showing how energy moves through a community. In a rain forest community, energy flows from the Sun to a rain forest tree, a producer. The tree uses the energy and grows, producing leaves and other plant structures. Energy moves to consumers, such as the sloth that eats the leaves of the tree, and then to the eagle that eats the sloth. When the eagle dies, detritivores, such as bacteria, feed on its body. That food chain can be written like this:

$$\text{Sun} \longrightarrow \text{leaves} \longrightarrow \text{sloth} \longrightarrow \text{eagle} \longrightarrow \text{bacteria}$$

A food chain shows only part of the energy flow in a community. A food web, like the one in **Figure 16,** shows many food chains within a community and how they overlap.

 Key Concept Check Identify a food chain in a community near your home. List the producers and consumers in your food chain.

Figure 16 Organisms in a rain forest community get their energy in different ways.

Visual Check List the members of two different food chains shown in the figure.

Food Web

Review   Personal Tutor

Relationships in Communities

The populations that make up a community interact with each other in a variety of ways. Some species have feeding relationships—they either eat or are eaten by another species. Some species interact with another species to get the food or shelter they need.

Predator-Prey Relationships

Hungry squirrel monkeys quarrel over a piece of fruit. They don't notice the harpy eagle above them. Suddenly, the eagle swoops down and grabs one of the monkeys in its talons. Harpy eagles and monkeys have a predator-prey relationship. The eagle, like other **predators,** hunts other animals for food. The hunted animals, such as the squirrel monkey or the lizard shown at the beginning of this lesson, are called prey.

As you read in Lesson 1, predators help prevent prey populations from growing too large for the carrying capacity of the ecosystem. The sand lizard, shown in **Figure 17,** is a predator in most of Europe. Like all predators, they often capture weak or injured individuals of a prey population. When the weak members of a population are removed, there are more resources available for the remaining members. This helps keep the prey population healthy.

 Reading Check Why are predators important to a prey population?

Figure 17 Sand lizards eat slugs, spiders, insects, fruits, and flowers.

Visual Check Which type of consumers are sand lizards?

Figure 18 Leaf-cutter ants cooperate while growing food. They work together and cut apart leaves. The ants carry the leaves to an underground nest. The ants' only food is a species of fungus that grows on the leaves.

Cooperative Relationships

The members of some populations work together in cooperative relationships for their survival, like the leaf-cutter ants shown in **Figure 18.** As you read in Lesson 1, meerkats cooperate with each other and raise young and watch for predators. Squirrel monkeys benefit in a similar way by living in groups. They cooperate as they hunt for food and watch for danger.

Symbiotic Relationships

Some species have such close relationships that they are almost always found living together. *A close, long-term relationship between two species that usually involves an exchange of food or energy is called* **symbiosis** (sihm bee OH sus). There are three types of symbiosis—mutualism, commensalism, and parasitism.

Mutualism Boxer crabs and sea anemones share a mutualistic partnership, as shown in **Figure 19.** *A symbiotic relationship in which both partners benefit is called* **mutualism.** Boxer crabs and sea anemones live in tropical coral reef communities. The crabs carry sea anemones in their claws. The sea anemones have stinging cells that help the crabs fight off predators. The sea anemones eat leftovers from the crabs' meals.

◀ **Figure 19** Boxer crabs and sea anemones have a mutualistic relationship because both partners benefit from the relationship.

▲ **Figure 20** Epiphytes and trees share a commensal relationship.

Figure 21 Hunting wasps are examples of parasites. The larvae use the paralyzed spider as food while they mature. ▼

Commensalism *A symbiotic relationship that benefits one species but does not harm or benefit the other is* **commensalism.** Plants called epiphytes (EH puh fites), shown in **Figure 20,** grow on the trunks of trees and other objects. The roots of an epiphyte anchor it to the object. The plant's nutrients are absorbed from the air. Epiphytes benefit from attaching to tree trunks by getting more living space and sunlight. The trees are neither helped nor harmed by the plants. Orchids are another example of epiphytes that have commensal relationships with trees.

Parasitism *A symbiotic relationship that benefits one species and harms the other is* **parasitism.** The species that benefits is the parasite. The species that is harmed is the host. Heartworms, tapeworms, fleas, and lice are parasites that feed on a host organism, such as a human or a dog. The parasites benefit by getting food. The host usually is not killed, but it can be weakened. For example, heartworms in a dog can cause the heart to work harder. Eventually, the heart can fail, killing the host. Other common parasites include the fungi that cause ringworm and toenail fungus. The fungi that cause these ailments feed on keratin (KER ah tihn), a protein in skin and nails.

The larvae of the hunting wasp is another example of a parasite. The female wasp, shown in **Figure 21,** stings a spider to paralyze it. Then she lays eggs in its body. When the eggs hatch into larvae, they eat the paralyzed spider's body. Another example of parasitism is the strangler fig. The seeds of the strangler fig sprout on the branches of a host tree. The young strangler fig sends roots into the tree and down into the ground below. The host tree provides the fig with nutrients and a trunk for support. Strangler figs grow fast and they can kill a host tree.

 Key Concept Check List five ways species in a community interact.

Lesson 3 Review

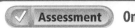

Visual Summary

Each organism in a community has its own habitat and niche within the ecosystem.

Within a community, each organism must obtain energy for life processes. Some organisms are producers and some are consumers.

Some organisms have cooperative relationships and some have symbiotic relationships. The hunting wasp and spider have a symbiotic relationship.

FOLDABLES®

Use your lesson Foldable to review the lesson. Save your Foldable for the project at the end of the chapter.

What do you think NOW?

You first read the statements below at the beginning of the chapter.

5. No more than two species can live in the same habitat.

6. A cow is a producer because it produces food for other organisms.

Did you change your mind about whether you agree or disagree with the statements? Rewrite any false statements to make them true.

Use Vocabulary

1. **Define** *symbiosis*.

2. **Distinguish** between producers and consumers.

Understand Key Concepts

3. **Explain** how energy from the Sun flows through a rain forest community.

4. **Compare and contrast** predator-prey relationships and cooperative relationships.

5. A shrimp removes and eats the parasites from the gills of a fish. The fish stays healthier because the parasites are removed. This relationship is
   A. commensalism. C. mutualism.
   B. competition.   D. parasitism.

Interpret Graphics

6. **Organize Information** Copy and fill in the table below with details about the three different types of symbiosis.

7. **Identify** the type of diagram shown below and explain what it means.

Critical Thinking

8. **Predict** what could happen to a population of ants if anteaters, a predator of the ants, disappeared.

9. **Decide** which type of symbiosis this is: Bacteria live in the skin under the eyes of deep-sea fish. The bacteria give off light that helps the fish find food. The bacteria get food from the fish.

Materials

symbiosis
cards

How can you model a symbiotic relationship?

As you read earlier, organisms in communities can have many different types of relationships. Symbiotic relationships occur when two organisms live in direct contact and form a relationship. Symbiotic relationships include mutualism, commensalism, and parasitism. Although communities around the world have symbiotic relationships, coral reef communities often include all three types of symbiosis. Many of the organisms in these communities, such as clownfish, sea anemones, and even microscopic copepods, have some type of symbiotic relationship. In this lab, you will research and model one type of symbiosis in a coral reef community.

Question

How do you model a symbiotic relationship and determine its type?

Procedure

1. Read and complete a lab safety form.

2. Get a card from your teacher with the name of an organism that has a symbiotic relationship. Find your partner(s) in the symbiotic relationship.

3. With your partner, brainstorm what type of symbiotic relationship your organism and your partner's organism might have. List and explain your choice(s) in your Science Journal.

4 Using your library and reference books, research your organism with your partner.

5 Develop a visual presentation, such as a skit, a slide presentation, or a series of posters with your partner showing how your symbiotic relationship works and how your organisms interact with other members of the community.

6 Show your presentation to the class.

Analyze and Conclude

7 **Identify** What type of symbiotic relationship did your organism have? What was your organism's role in the relationship?

8 **Compare** How would your organism interact in the community if its partner were not present?

9 **Contrast** What other organisms in a coral reef community have the same type of symbiotic relationship as your organism? If none, explain why.

10 **The Big Idea** How did your organism interact with other members of its population and community?

Communicate Your Results

Make a poster illustrating all the symbiotic relationships you and your classmates studied. Determine what type of relationship each example had. Identify which organisms are hosts, if any.

Inquiry Extension

All of the organisms your class studied are part of the coral reef ecosystem. Create a food web showing how the organisms obtained energy.

Lab Tips

☑ Think about your organism's niche in the ecosystem.

☑ Carefully select resources for accuracy.

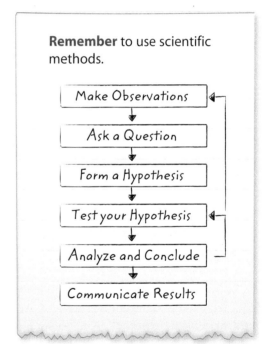

Remember to use scientific methods.

- Make Observations
- Ask a Question
- Form a Hypothesis
- Test your Hypothesis
- Analyze and Conclude
- Communicate Results

Chapter 21 Study Guide

THE BIG IDEA

A community contains many populations that interact in their energy roles and in their competition for resources. Populations can increase, decrease, and move, affecting the community.

Key Concepts Summary 🔑

Vocabulary

Lesson 1: Populations

- A **population** is all the organisms of the same species that live in the same area at the same time.
- Population sizes vary due to **limiting factors** such as environmental factors and available resources.
- Population size usually does not exceed the **carrying capacity** of the ecosystem.

biosphere p. 741
community p. 742
population p. 742
competition p. 743
limiting factor p. 743
population density p. 744
biotic potential p. 744
carrying capacity p. 745

Lesson 2: Changing Populations

- Populations of living things can increase, decrease, or move.
- Populations can decrease until they are threatened, endangered, or extinct.
- Human population size is affected by the same three factors as other populations—**birthrate**, **death rate**, and movement.

birthrate p. 749
death rate p. 749
extinct species p. 751
endangered species p. 751
threatened species p. 751
migration p. 752

Lesson 3: Communities

- A community is all the populations of different species that live together in the same area at the same time.
- The place within an ecosystem where an organism lives is its **habitat** and what an organism does in its habitat to survive is its **niche.**
- Three types of relationships within a community are predator-prey, cooperative, and symbiotic.

habitat p. 759
niche p. 759
producer p. 760
consumer p. 760
symbiosis p. 763
mutualism p. 763
commensalism p. 764
parasitism p. 764

Study Guide

 **Review**
- **Personal Tutor**
- **Vocabulary eGames**
- **Vocabulary eFlashcards**

FOLDABLES®

Assemble your lesson Foldables as shown to make a Chapter Project. Use the project to review what you have learned in this chapter.

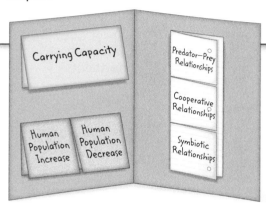

Use Vocabulary

1 The struggle in a community for the same resources is _____.

2 The part of Earth that supports life is the _____.

3 The instinctive movement of a population is _____.

4 A(n) _____ species is one at risk of becoming endangered.

5 A(n) _____ is an organism that gets energy from the environment.

6 The largest number of offspring that can be produced when there are no limiting factors is the _____.

Link Vocabulary and Key Concepts

 **Concepts in Motion** Interactive Concept Map

Copy this concept map, and then use vocabulary terms from the previous page to complete the concept map.

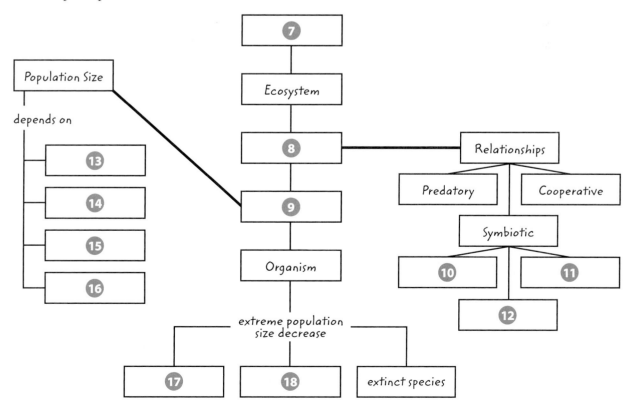

Understand Key Concepts 🔑

1 What does the line indicated by the red arrow in the graph below represent?

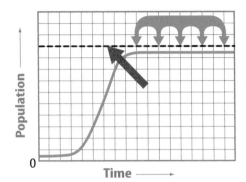

A. competition
B. biotic potential
C. carrying capacity
D. limiting factors

2 The need for organisms to rely on the same resources causes

A. competition.
B. biotic potential.
C. carrying capacity.
D. population growth.

3 The number of organisms in a specific area is

A. a community.
B. the carrying capacity.
C. the population density.
D. the population growth.

4 The number of robins that hatch in a year is the population's

A. biotic potential.
B. birthrate.
C. carrying capacity.
D. exponential growth.

5 A robin population that reaches its biotic potential probably shows

A. exponential growth.
B. low growth.
C. negative growth.
D. no growth.

6 An organism that uses sunlight to make food molecules is a(n)

A. carnivore.
B. consumer.
C. herbivore.
D. producer.

7 Which is NOT part of Earth's biosphere?

A. low atmosphere
B. surface of the Moon
C. bottom of the Pacific ocean
D. North American continent

8 Which is a limiting factor for a cottontail rabbit population on the prairie in Oklahoma?

A. a large amount of food
B. a large amount of shelter space
C. an abundance of coyotes in the area
D. an unpolluted river in the ecosystem

9 Which factor does NOT normally affect human population size?

A. birthrate
B. death rate
C. population movement
D. lack of resources

10 What type of overall population change is shown below?

A. immigration
B. migration
C. population decrease
D. population increase

Critical Thinking

11 **Select and draw** three food chains from the food web shown.

12 **Give an Example** What problems might result from overpopulation of pigeons in a city park?

13 **Describe** What are some possible solutions that a city might use to solve a pigeon overpopulation problem?

14 **Decide** Would sample counting or capture-mark-and-release at a specified time and place be the best method for measuring each of these populations: birds, whales, bluebonnet flowers, and oak trees?

15 **Compare and contrast** the feeding habits of carnivores, omnivores, and producers.

16 **Classify** Decide whether each of these relationships is mutualism, commensalism, or parasitism.

- Butterfly pollinates flower while drinking nectar.

- Tapeworm feeds on contents of dog's intestines.

- Fish finds shelter in coral reef.

17 **Draw** a food web that describes energy flow in this community. Insects eat leaves. Spiders eat insects. Birds eat insects and spiders. Frogs eat insects. Birds eat frogs.

Writing in Science

18 **Write** a two-page story that explains how an imaginary population becomes threatened with extinction.

REVIEW THE BIG IDEA

19 Describe three different types of relationships in a community between two different populations of organisms.

20 Do you think this large number of pigeons affects other organisms in the community? Explain your answer.

Math Skills

Review

Math Practice

Use Graphs
Use the graph to answer the questions.

21 During what range of years did the population change the least?

22 The dotted line represents a prediction. What does it predict about population growth beyond the present time?

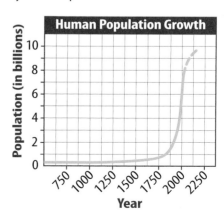

Standardized Test Practice

Record your answers on the answer sheet provided by your teacher or on a sheet of paper.

Multiple Choice

1 Which is defined as the demand for resources in short supply in a community?

   A biotic potential

   B competition

   C density

   D limiting factor

Use the diagram below to answer questions 2 and 3.

2 In the diagram above, which number represents an ecosystem?

   A 1

   B 2

   C 3

   D 4

3 According to the arrows, how are the elements of the diagram organized?

   A endangered to overpopulated

   B farthest to nearest

   C largest to smallest

   D nonliving to living

4 Which is NOT a possible result of overpopulation?

   A damage to soil

   B increased carrying capacity

   C loss of trees and plants

   D more competition

Use the diagram below to answer question 5.

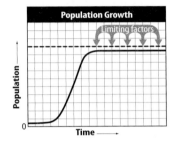

5 What does the dashed line in the diagram represent?

   A biotic potential

   B carrying capacity

   C overpopulation

   D population density

6 What does a population undergo when it has no limiting factors?

   A exponential growth

   B extinction

   C migration

   D population movement

7 What is the term for all species living in the same ecosystem at the same time?

   A biosphere

   B community

   C habitat

   D population

Use the diagram below to answer question 8.

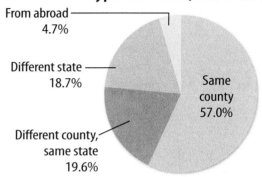

Types of Moves, 2004–2005

From abroad — 4.7%

Different state — 18.7%

Same county 57.0%

Different county, same state 19.6%

Source: U.S. Census Bureau, Current Population Survey, 2005 Annual Social and Economic Supplement.

8 According to the diagram, about how many of those who moved remained within the same state?

  A 20 percent

  B 39 percent

  C 57 percent

  D 77 percent

9 Which type of relationship includes mutualism and parasitism?

  A competition

  B cooperation

  C predation

  D symbiosis

10 Which is a population?

  A all meerkats in a refuge

  B all the types of birds in a forest

  C all the types of cats in a zoo

  D all the types of insects in a swamp

Constructed Response

Use the table below to answer questions 11 and 12.

Growth	Decline

11 In the table, list four factors that contribute to human population growth and four factors that lead to population decline.

12 Select one factor from each column in the table above. Which has the greatest effect on human population today? Explain your reasoning.

13 Describe a negative consequence of human population growth. How can humans minimize the effect of this change?

Use the diagram below to answer question 14.

sunlight → grasses → antelope → lion → bacteria

14 Explain how each organism in the food chain above gets energy. How might the other organisms be affected if the antelope population declines due to disease?

NEED EXTRA HELP?														
If You Missed Question...	1	2	3	4	5	6	7	8	9	10	11	12	13	14
Go to Lesson...	1	1	1	1	1	2	1	2	3	1	2	2	2	3

Biomes and Ecosystems

THE BIG IDEA

How do Earth's biomes and ecosystems differ?

Inquiry **Modern Art?**

Although it might look like a piece of art, this structure was designed to replicate several ecosystems. When Biosphere 2 was built in the 1980s near Tucson, Arizona, it included a rain forest, a desert, a grassland, a coral reef, and a wetland. Today, it is used mostly for research and education.

- How realistic do you think Biosphere 2 is?

- Is it possible to make artificial environments as complex as those in nature?

- How do Earth's biomes and ecosystems differ?

Get Ready to Read

What do you think?

Before you read, decide if you agree or disagree with each of these statements. As you read this chapter, see if you change your mind about any of the statements.

1 Deserts can be cold.

2 There are no rain forests outside the tropics.

3 Estuaries do not protect coastal areas from erosion.

4 Animals form coral reefs.

5 An ecosystem never changes.

6 Nothing grows in the area where a volcano has erupted.

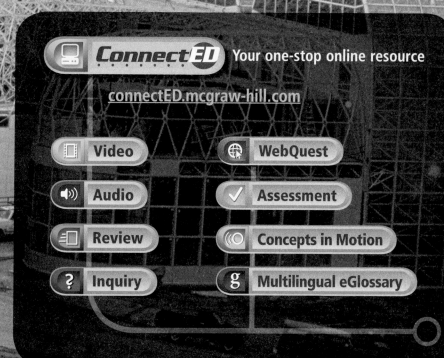

Connect ED Your one-stop online resource

connectED.mcgraw-hill.com

□ Video

⊕ WebQuest

◀)) Audio

✓ Assessment

▤ Review

◉ Concepts in Motion

? Inquiry

g Multilingual eGlossary

◀))

Lesson 1

Reading Guide

Key Concepts 🔑
ESSENTIAL QUESTIONS

- How do Earth's land biomes differ?
- How do humans impact land biomes?

Vocabulary

biome p. 777

desert p. 778

grassland p. 779

temperate p. 781

taiga p. 783

tundra p. 783

 Multilingual eGlossary

📺 Video BrainPOP®

Land Biomes

Inquiry Plant or Animal?

Believe it or not, this is a flower. One of the largest flowers in the world, *Rafflesia* (ruh FLEE zhuh), grows naturally in the tropical rain forests of southeast Asia. What do you think would happen if you planted a seed from this plant in a desert? Would it survive?

Launch Lab

10 minutes

What is the climate in China?

Beijing, China, and New York, New York, are about the same distance from the equator but on opposite sides of Earth. How do temperature and rainfall compare for these two cities?

1. Locate Beijing and New York on a world map.

2. Copy the table to the right in your Science Journal. From the data and charts provided, find and record the average high and low temperatures in January and in June for each city.

3. Record the average rainfall in January and in June for each city.

Think About This

1. What are the temperature and rainfall ranges for each city?

2. 🔑 **Key Concept** How do you think the climates of these cities differ year-round?

High Temperature (°C)	January	June
Beijing		
New York		
Low Temperature (°C)	**January**	**June**
Beijing		
New York		
Rainfall (mm)	**January**	**June**
Beijing		
New York		

Land Ecosystems and Biomes

When you go outside, you might notice people, grass, flowers, birds, and insects. You also are probably aware of nonliving things, such as air, sunlight, and water. The living or once-living parts of an environment are the biotic parts. The nonliving parts that the living parts need to survive are the abiotic parts. The biotic and abiotic parts of an environment together make up an ecosystem.

Earth's continents have many different ecosystems, from deserts to rain forests. Scientists classify similar ecosystems in large geographic areas as biomes. *A biome is a geographic area on Earth that contains ecosystems with similar biotic and abiotic features.* As shown in **Figure 1,** Earth has seven major land biomes. Areas classified as the same biome have similar climates and organisms.

Figure 1 Earth contains seven major biomes.

Concepts in Motion Animation

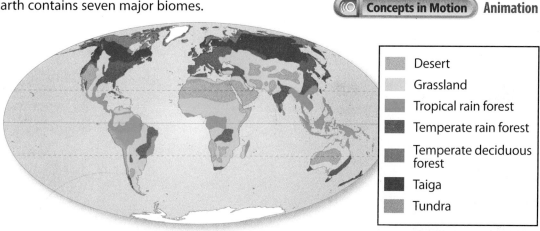

Legend:
- Desert
- Grassland
- Tropical rain forest
- Temperate rain forest
- Temperate deciduous forest
- Taiga
- Tundra

How hot is sand?

If you have ever walked barefoot on a sandy beach on a sunny day, you know how hot sand can be. But how hot is the sand below the surface?

1. Read and complete a lab safety form.

2. Position a **desk lamp** over a **container** of **sand** that is at least 7 cm deep.

3. Place one **thermometer** on the surface of the sand and bury the tip of another **thermometer** about 5 cm below the surface. Record the temperature on each thermometer in your Science Journal.

4. Turn on the lamp and record the temperatures again after 10 minutes.

Analyze and Conclude

1. **Describe** the temperatures of the sand at the surface and below the surface.

2. **Predict** what would happen to the temperature of the sand at night.

3. 🔑 **Key Concept** Desert soil contains a high percentage of sand. Based on your results, predict ways in which species are adapted to living in an environment where the soil is mostly sand.

Desert Biome   Woodpeckers

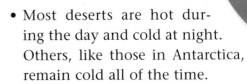

Deserts *are biomes that receive very little rain.* They are on nearly every continent and are Earth's driest ecosystems.

- Most deserts are hot during the day and cold at night. Others, like those in Antarctica, remain cold all of the time.

- Rainwater drains away quickly because of thin, porous soil. Large patches of ground are bare.

Biodiversity

- Animals include lizards, bats, woodpeckers, and snakes. Most animals avoid activity during the hottest parts of the day.

- Plants include spiny cactus and thorny shrubs. Shallow roots absorb water quickly. Some plants have accordion-like stems that expand and store water. Small leaves or spines reduce the loss of water.

Human Impact

- Cities, farms, and recreational areas in deserts use valuable water.

- Desert plants grow slowly. When they are damaged by people or livestock, recovery takes many years.

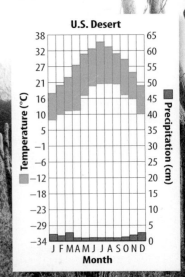

U.S. Desert

Temperature (°C): 38, 32, 27, 21, 16, 10, 5, −1, −6, −12, −18, −23, −29, −34

Precipitation (cm): 65, 60, 55, 50, 45, 40, 35, 30, 25, 20, 15, 10, 5, 0

Month: J F M A M J J A S O N D

Black-footed ferret

Grassland *biomes are areas where grasses are the dominant plants.* Also called prairies, savannas, and meadows, grasslands are the world's "breadbaskets." Wheat, corn, oats, rye, barley, and other important cereal crops are grasses. They grow well in these areas.

- Grasslands have a wet and a dry season.

- Deep, fertile soil supports plant growth.

- Grass roots form a thick mass, called sod, which helps soil absorb and hold water during periods of drought.

Reading Check Why are grasslands called "breadbaskets"?

Biodiversity

- Trees grow along moist banks of streams and rivers. Wildflowers bloom during the wet season.

- In North America, large herbivores, such as bison and elk, graze here. Insects, birds, rabbits, prairie dogs, and snakes find shelter in the grasses.

- Predators in North American grasslands include hawks, ferrets, coyotes, and wolves.

- African savannas are grasslands that contain giraffes, zebras, and lions. Australian grasslands are home to kangaroos, wallabies, and wild dogs.

Human Impact

- People plow large areas of grassland to raise cereal crops. This reduces habitat for wild species.

- Because of hunting and loss of habitat, large herbivores—such as bison—are now uncommon in many grasslands.

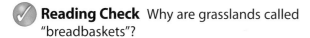

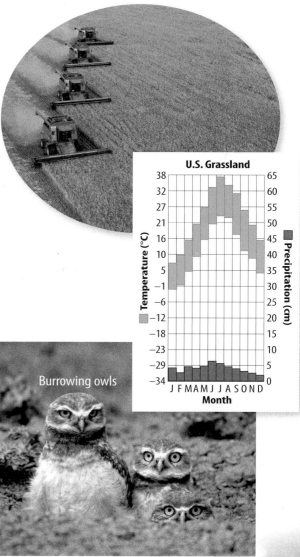
Burrowing owls

U.S. Grassland

Tropical Rain Forest Biome

Ocelot

Toucan

The forests that grow near the equator are called tropical rain forests. These forests receive large amounts of rain and have dense growths of tall, leafy trees.

- Weather is warm and wet year-round.

- The soil is shallow and easily washed away by rain.

- Less than 1 percent of the sunlight that reaches the top of forest trees also reaches the forest floor.

- Half of Earth's species live in tropical rain forests. Most live in the canopy—the uppermost part of the forest.

✓ **Reading Check** Where do most organisms live in a tropical rain forest?

Biodiversity

- Few plants live on the dark forest floor.

- Vines climb the trunks of tall trees.

- Mosses, ferns, and orchids live on branches in the canopy.

- Insects make up the largest group of tropical animals. They include beetles, termites, ants, bees, and butterflies.

- Larger animals include parrots, toucans, snakes, frogs, flying squirrels, fruit bats, monkeys, jaguars, and ocelots.

Human Impact

- People have cleared more than half of Earth's tropical rain forests for lumber, farms, and ranches. Poor soil does not support rapid growth of new trees in cleared areas.

- Some organizations are working to encourage people to use less wood harvested from rain forests.

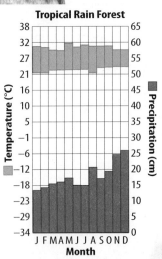

Tropical Rain Forest

Temperature (°C): 38, 32, 27, 21, 16, 10, 5, -1, -6, -12, -18, -23, -29, -34

Precipitation (cm): 65, 60, 55, 50, 45, 40, 35, 30, 25, 20, 15, 10, 5, 0

Month: J F M A M J J A S O N D

Temperate Rain Forest Biome

Regions of Earth between the tropics and the polar circles are **temperate** *regions.* Temperate regions have relatively mild climates with distinct seasons. Several biomes are in temperate regions, including rain forests. Temperate rain forests are moist ecosystems mostly in coastal areas. They are not as warm as tropical rain forests.

Elk

• Winters are mild and rainy.

• Summers are cool and foggy.

• Soil is rich and moist.

Biodiversity

• Forests are dominated by spruce, hemlock, cedar, fir, and redwood trees, which can grow very large and tall.

• Fungi, ferns, mosses, vines, and small flowering plants grow on the moist forest floor.

• Animals include mosquitoes, butterflies, frogs, salamanders, woodpeckers, owls, eagles, chipmunks, raccoons, deer, elk, bears, foxes, and cougars.

Human Impact

• Temperate rain forest trees are a source of lumber. Logging can destroy the habitat of forest species.

• Rich soil enables cut forests to grow back. Tree farms help provide lumber without destroying habitat.

 Key Concept Check In what ways do humans affect temperate rain forests?

FOLDABLES

Use a sheet of paper to make a horizontal two-tab book. Record what you learn about desert and temperate rain forest biomes under the tabs, and use the information to compare and contrast these biomes.

Desert Biome | Temperate Rain Forest Biome

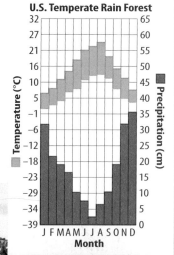

U.S. Temperate Rain Forest

Temperate Deciduous Forest Biome

Temperate deciduous forests grow in temperate regions where winter and summer climates have more variation than those in temperate rain forests. These forests are the most common forest ecosystems in the United States. They contain mostly deciduous trees, which lose their leaves in the fall.

- Winter temperatures are often below freezing. Snow is common.

- Summers are hot and humid.

- Soil is rich in nutrients and supports a large amount of diverse plant growth.

Biodiversity

- Most plants, such as maples, oaks, birches, and other deciduous trees, stop growing during the winter and begin growing again in the spring.

- Animals include snakes, ants, butterflies, birds, raccoons, opossums, and foxes.

- Some animals, including chipmunks and bats, spend the winter in hibernation.

- Many birds and some butterflies, such as the monarch, migrate to warmer climates for the winter.

Human Impact

Over the past several hundred years, humans have cleared thousands of acres of Earth's deciduous forests for farms and cities. Today, much of the clearing has stopped and some forests have regrown.

 Key Concept Check How are temperate deciduous rain forests different from temperate rain forests?

U.S. Temperate Deciduous Forest

Temperature (°C) / Precipitation (cm) versus Month (J F M A M J J A S O N D)

Red fox

Taiga Biome

A **taiga** (TI guh) *is a forest biome consisting mostly of cone-bearing evergreen trees.* The taiga biome exists only in the northern hemisphere. It occupies more space on Earth's continents than any other biome.

- Winters are long, cold, and snowy. Summers are short, warm, and moist.
- Soil is thin and acidic.

Biodiversity

- Evergreen trees, such as spruce, pine, and fir, are thin and shed snow easily.
- Animals include owls, mice, moose, bears, and other cold-adapted species.
- Abundant insects in summer attract many birds, which migrate south in winter.

Human Impact

- Tree harvesting reduces taiga habitat.

Brown bear

Tundra Biome

A **tundra** (TUN druh) *biome is cold, dry, and treeless.* Most tundra is south of the North Pole, but it also exists in mountainous areas at high altitudes.

- Winters are long, dark, and freezing; summers are short and cool; the growing season is only 50–60 days.
- Permafrost—a layer of permanently frozen soil—prevents deep root growth.

Biodiversity

- Plants include shallow-rooted mosses, lichens, and grasses.
- Many animals hibernate or migrate south during winter. Few animals, including lemmings, live in tundras year-round.

Human Impact

- Drilling for oil and gas can interrupt migration patterns.

Lemming

Lesson 1 Review

Visual Summary

Earth has seven major land biomes, ranging from hot, dry deserts to cold, forested taigas.

Half of Earth's species live in rain forest biomes.

Temperate deciduous forests are the most common forest biome in the United States.

FOLDABLES®

Use your lesson Foldable to review the lesson. Save your Foldable for the project at the end of the chapter.

What do you think NOW?

You first read the statements below at the beginning of the chapter.

1. Deserts can be cold.

2. There are no rain forests outside the tropics.

Did you change your mind about whether you agree or disagree with the statements? Rewrite any false statements to make them true.

Use Vocabulary

1 **Define** *biome* using your own words.

2 **Distinguish** between tropical rain forests and temperate rain forests.

3 A cold, treeless biome is a(n) _____.

Understand Key Concepts

4 **Explain** why tundra soil cannot support the growth of trees.

5 **Give examples** of how plants and animals adapt to temperate deciduous ecosystems.

Interpret Graphics

6 **Determine** What is the average annual rainfall for the biome represented by the chart to the right?

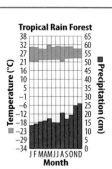

7 **Summarize Information** Copy the graphic organizer below and fill it in with animals and plants of the biome you live in.

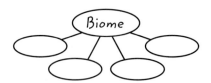

Critical Thinking

8 **Plan** an enclosed zoo exhibit for a desert ecosystem. What abiotic factors should you consider?

9 **Recommend** one or more actions people can take to reduce habitat loss in tropical and taiga forests.

Which biome is it?

Materials

biome data

You have read about the major land biomes found on Earth. Within each biome are ecosystems with similar biotic and abiotic factors. In this lab, you will **interpret data** describing a particular area on Earth to identify which biome it belongs to.

Learn It

Scientists collect and present data in a variety of forms, including graphs and tables. In this activity, you will interpret data in a graph and apply the information to the ideas you learned in the lesson.

Try It

1. Examine the temperature and precipitation data in the graph given you by your teacher.

2. Create a table from these data in your Science Journal. Calculate the average temperature and precipitation during the winter and the summer.

3. Examine the image of the biome and identify some plants and animals in the image.

4. Compare your data to the information on land biomes presented in Lesson 1. Which biome is the most similar?

Apply It

5. Which land biome did your data come from? Why did you choose this biome?

6. Are the data in your graph identical to the data in the graph of the biome in Lesson 1 to which it belongs? Why or why not?

7. Describe this biome. What do you think your biome will be like six months from now?

8. 🔑 **Key Concept** How might humans affect the organisms in your biome?

Aquatic Ecosystems

Reading Guide

Key Concept 🔑
ESSENTIAL QUESTIONS

- How do Earth's aquatic ecosystems differ?
- How do humans impact aquatic ecosystems?

Vocabulary

salinity p. 787

wetland p. 790

estuary p. 791

intertidal zone p. 793

coral reef p. 793

g Multilingual eGlossary

Inquiry Floating Trees?

These plants, called mangroves, are one of the few types of plants that grow in salt water. They usually live along ocean coastlines in tropical ecosystems. What other organisms do you think live near mangroves?

What happens when rivers and oceans mix?

Freshwater and saltwater ecosystems have different characteristics. What happens in areas where freshwater rivers and streams flow into oceans?

1. Read and complete a lab safety form.

2. In a **plastic tub**, add 100 g of **salt** to 2 L of water. Stir with a **long-handled spoon** until the salt dissolves.

3. In another **container**, add 5 drops of **blue food coloring** to 1 L of water. Gently pour the colored water into one corner of the plastic tub. Observe how the color of the water changes in the tub.

4. Observe the tub again in 5 minutes.

Think About This

1. What bodies of water do the containers represent?

2. What happened to the water in the tub after 5 minutes? What do you think happens to the salt content of the water?

3. **Key Concept** How do you think the biodiversity of rivers and oceans differ? What organisms do you think might live at the place where the two meet?

Aquatic Ecosystems

If you've ever spent time near an ocean, a river, or another body of water, you might know that water is full of life. There are four major types of water, or aquatic, ecosystems: freshwater, wetland, estuary, and ocean. Each type of ecosystem contains a unique variety of organisms. Whales, dolphins, and corals live only in ocean ecosystems. Catfish and trout live only in freshwater ecosystems. Many other organisms that do not live under water, such as birds and seals, also depend on aquatic ecosystems for food and shelter.

Important abiotic factors in aquatic ecosystems include temperature, sunlight, and dissolved oxygen gas. Aquatic species have adaptations that enable them to use the oxygen in water. The gills of a fish separate oxygen from water and move it into the fish's bloodstream. Mangrove plants, pictured on the previous page, take in oxygen through small pores in their leaves and roots.

Salinity (say LIH nuh tee) is another important abiotic factor in aquatic ecosystems. **Salinity** *is the amount of salt dissolved in water.* Water in saltwater ecosystems has high salinity compared to water in freshwater ecosystems, which contains little salt.

Math Skills

Use Proportions

Salinity is measured in parts per thousand (PPT). One PPT water contains 1 g salt and 1,000 g water. Use proportions to calculate salinity. What is the salinity of 100 g of water with 3.5 g of salt?

$$\frac{3.5 \text{ g salt}}{100 \text{ g seawater}} = \frac{x \text{ g salt}}{1,000 \text{ g seawater}}$$

$$100\,x = 3500$$

$$x = \frac{3500}{100} = 35 \text{ PPT}$$

Practice

A sample contains 0.1895 g of salt per 50 g of seawater. What is its salinity?

 Review

- **Math Practice**
- **Personal Tutor**

Great Blue Heron

Freshwater ecosystems include streams, rivers, ponds, and lakes. Streams are usually narrow, shallow, and fast-flowing. Rivers are larger, deeper, and flow more slowly.

- Streams form from underground sources of water, such as springs or from runoff from rain and melting snow.

- Stream water is often clear. Soil particles are quickly washed downstream.

- Oxygen levels in streams are high because air mixes into the water as it splashes over rocks.

- Rivers form when streams flow together.

- Soil that washes into a river from streams or nearby land can make river water muddy. Soil also introduces nutrients, such as nitrogen, into rivers.

- Slow-moving river water has higher levels of nutrients and lower levels of dissolved oxygen than fast-moving water.

Biodiversity

- Willows, cottonwoods, and other water-loving plants grow along streams and on riverbanks.

- Species adapted to fast-moving water include trout, salmon, crayfish, and many insects.

- Species adapted to slow-moving water include snails and catfish.

Stonefly
larva

Salmon

Human Impact

- People take water from streams and rivers for drinking, laundry, bathing, crop irrigation, and industrial purposes.

- Hydroelectric plants use the energy in flowing water to generate electricity. Dams stop the water's flow.

- Runoff from cities, industries, and farms is a source of pollution.

Freshwater: Ponds and Lakes

Ponds and lakes contain freshwater that is not flowing downhill. These bodies of water form in low areas on land.

- Ponds are shallow and warm.
- Sunlight reaches the bottom of most ponds.
- Pond water is often high in nutrients.
- Lakes are larger and deeper than ponds.
- Sunlight penetrates into the top few feet of lake water. Deeper water is dark and cold.

Biodiversity

- Plants surround ponds and lake shores.
- Surface water in ponds and lakes contains plants, algae, and microscopic organisms that use sunlight for photosynthesis.
- Organisms living in shallow water near shorelines include cattails, reeds, insects, crayfish, frogs, fish, and turtles.
- Fewer organisms live in the deeper, colder water of lakes where there is little sunlight.
- Lake fish include perch, trout, bass, and walleye.

 Reading Check Why do few organisms live in the deep water of lakes?

Human Impact

- Humans fill in ponds and lakes with sediment to create land for houses and other structures.
- Runoff from farms, gardens, and roads washes pollutants into ponds and lakes, disrupting food webs.

 Key Concept Check How do ponds and lakes differ?

Smallmouth bass

Common loon

Some types of aquatic ecosystems have mostly shallow water. **Wetlands** *are aquatic ecosystems that have a thin layer of water covering soil that is wet most of the time.* Wetlands contain freshwater, salt water, or both. They are among Earth's most fertile ecosystems.

- Freshwater wetlands form at the edges of lakes and ponds and in low areas on land. Saltwater wetlands form along ocean coasts.

- Nutrient levels and biodiversity are high.

- Wetlands trap sediments and purify water. Plants and microscopic organisms filter out pollution and waste materials.

Biodiversity

- Water-tolerant plants include grasses and cattails. Few trees live in saltwater wetlands. Trees in freshwater wetlands include cottonwoods, willows, and swamp oaks.

- Insects are abundant and include flies, mosquitoes, dragonflies, and butterflies.

- More than one-third of North American bird species, including ducks, geese, herons, loons, warblers, and egrets, use wetlands for nesting and feeding.

- Other animals that depend on wetlands for food and breeding grounds include alligators, turtles, frogs, snakes, salamanders, muskrats, and beavers.

Human Impact

- In the past, many people considered wetlands as unimportant environments. Water was drained away to build homes and roads and to raise crops.

- Today, many wetlands are being preserved, and drained wetlands are being restored.

 Key Concept Check How do humans impact wetlands?

Estuaries

Estuaries (ES chuh wer eez) *are regions along coastlines where streams or rivers flow into a body of salt water.* Most estuaries form along coastlines, where freshwater in rivers meets salt water in oceans. Estuary ecosystems have varying degrees of salinity.

- Salinity depends on rainfall, the amount of freshwater flowing from land, and the amount of salt water pushed in by tides.

- Estuaries help protect coastal land from flooding and erosion. Like wetlands, estuaries purify water and filter out pollution.

- Nutrient levels and biodiversity are high.

Biodiversity

- Plants that grow in salt water include mangroves, pickleweeds, and seagrasses.

- Animals include worms, snails, and many species that people use for food, including oysters, shrimp, crabs, and clams.

- Striped bass, salmon, flounder, and many other ocean fish lay their eggs in estuaries.

- Many species of birds depend on estuaries for breeding, nesting, and feeding.

Human Impact

- Large portions of estuaries have been filled with soil to make land for roads and buildings.

- Destruction of estuaries reduces habitat for estuary species and exposes the coastline to flooding and storm damage.

WORD ORIGIN

estuary

from Latin *aestuarium*, means "a tidal marsh or opening."

FOLDABLES®

Make a horizontal two-tab book and label it as shown. Use it to compare how biodiversity and human impact differ in wetlands and estuaries.

Wetlands | Estuaries

Harvest mouse

Sunlit zone

200 m

Twilight zone

Continental shelf

1,000 m

Dark zone

3,800 m

Seafloor

Jellyfish

Fur seal

Most of Earth's surface is covered by ocean water with high salinity. The oceans contain different types of ecosystems. If you took a boat trip several kilometers out to sea, you would be in the open ocean—one type of ocean ecosystem. The open ocean extends from the steep edges of continental shelves to the deepest parts of the ocean. The amount of light in the water depends on depth.

- Photosynthesis can take place only in the uppermost, or sunlit, zone. Very little sunlight reaches the twilight zone. None reaches the deepest water, known as the dark zone.

- Decaying matter and nutrients float down from the sunlit zone, through the twilight and dark zones, to the seafloor.

Biodiversity

- Microscopic algae and other producers in the sunlit zone form the base of most ocean food chains. Other organisms living in the sunlit zone are jellyfish, tuna, mackerel, and dolphins.

- Many species of fish stay in the twilight zone during the day and swim to the sunlit zone at night to feed.

- Sea cucumbers, brittle stars, and other bottom-dwelling organisms feed on decaying matter that drifts down from above.

- Many organisms in the dark zone live near cracks in the seafloor where lava erupts and new seafloor forms.

Reading Check Which organisms are at the base of most ocean food chains?

Human Impact

- Overfishing threatens many ocean fish.

- Trash discarded from ocean vessels or washed into oceans from land is a source of pollution. Animals such as seals become tangled in plastic or mistake it for food.

Ocean: Coastal Oceans

Sea stars

Coastal oceans include several types of ecosystems, including continental shelves and intertidal zones. *The **intertidal zone** is the ocean shore between the lowest low tide and the highest high tide.*

- Sunlight reaches the bottom of shallow coastal ecosystems.

- Nutrients washed in from rivers and streams contribute to high biodiversity.

Biodiversity

- The coastal ocean is home to mussels, fish, crabs, sea stars, dolphins, and whales.

- Intertidal species have adaptations for surviving exposure to air during low tides and to heavy waves during high tides.

Human Impact

- Oil spills and other pollution harm coastal organisms.

Ocean: Coral Reefs

Another ocean ecosystem with high biodiversity is the coral reef. *A **coral reef** is an underwater structure made from outside skeletons of tiny, soft-bodied animals called coral.*

- Most coral reefs form in shallow tropical oceans.

- Coral reefs protect coastlines from storm damage and erosion.

Biodiversity

- Coral reefs provide food and shelter for many animals, including parrotfish, groupers, angelfish, eels, shrimp, crabs, scallops, clams, worms, and snails.

Human Impact

- Pollution, overfishing, and harvesting of coral threaten coral reefs.

inquiry MiniLab   15 minutes

How do ocean ecosystems differ?

Ocean ecosystems include open oceans, coastal oceans, and coral reefs—each one a unique environment with distinctive organisms.

1. Read and complete a lab safety form.

2. In a **large plastic tub**, use **rocks** and **sand** to make a structure representing an open ocean, a coastal ocean, or a coral reef.

3. Fill the tub with **water.**

4. Make waves by gently moving your hand back and forth in the water.

Analyze and Conclude

1. **Observe** What happened to your structure when you made waves? How might a hurricane affect the organisms that live in the ecosystem you modeled?

2. **Key Concept** Compare your results with results of those who modeled other ecosystems. Suggest what adaptations species might have in each ecosystem.

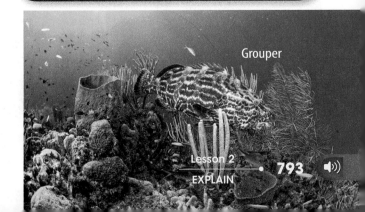

Grouper

Lesson 2 Review

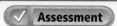

Visual Summary

Freshwater ecosystems include ponds and lakes.

Wetlands can be salt-water ecosystems or freshwater ecosystems.

Coral reefs and coastal ecosystems have high levels of biodiversity.

 FOLDABLES

Use your lesson Foldable to review the lesson. Save your Foldable for the project at the end of the chapter.

What do you think NOW?

You first read the statements below at the beginning of the chapter.

3. Estuaries do not protect coastal areas from erosion.

4. Animals form coral reefs.

Did you change your mind about whether you agree or disagree with the statements? Rewrite any false statements to make them true.

Use Vocabulary

1. **Define** the term *salinity*.

2. **Distinguish** between a wetland and an estuary.

3. An ocean ecosystem formed from the skeletons of animals is a(n) _____.

Understand Key Concepts

4. Which ecosystem contains both salt water and freshwater?
   - **A.** estuary
   - **B.** lake
   - **C.** pond
   - **D.** stream

5. **Describe** what might happen to a coastal area if its estuary were filled in to build houses.

Interpret Graphics

6. **Describe** Copy the drawing to the right and label the light zones. Describe characteristics of each zone.

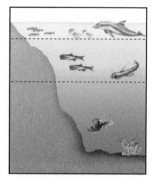

Critical Thinking

7. **Recommend** actions people might take to prevent pollutants from entering coastal ecosystems.

Math Skills

Review
— Math Practice —

8. The salinity of the Baltic Sea is about 10 PPT. What weight of salt is present in 2,000 g of its seawater?

Saving an Underwater Wilderness

A researcher takes a water sample from a marine reserve. ▼

How do scientists help protect coral reefs?

Pollution and human activities, such as mining and tourism, have damaged many ecosystems, including coral reefs. Scientists and conservation groups are working together to help protect and restore coral reefs and areas that surround them. One way is to create marine reserves where no fishing or collection of organisms is allowed.

A team of scientists, including marine ecologists Dr. Dan Brumbaugh and Kate Holmes from the American Museum of Natural History, are investigating how well reserves are working. These scientists compare how many fish of one species live both inside and outside reserves. Their results indicate that more species of fish and greater numbers of each species live inside reserves than outside—one sign that reefs in the area are improving.

Reef ecosystems do not have to be part of a reserve in order to improve, however. Scientists can work with local governments to find ways to limit damage to reef ecosystems. One way is to prevent overfishing by limiting the number of fish caught. Other ways include eliminating the use of destructive fishing practices that can harm reefs and reducing runoff from farms and factories.

By creating marine reserves, regulating fishing practices, and reducing runoff, humans can help reefs that were once in danger become healthy again.

Kate Holmes examines a coral reef. ▶

It's Your Turn

WRITE Write a persuasive essay describing why coral reefs are important habitats.

AMERICAN MUSEUM OF NATURAL HISTORY

Reading Guide

Key Concepts 🔑
ESSENTIAL QUESTIONS

- How do land ecosystems change over time?

- How do aquatic ecosystems change over time?

Vocabulary

ecological succession p. 797

climax community p. 797

pioneer species p. 798

eutrophication p. 800

 Multilingual eGlossary

 Video

- **Science Video**
- **What's Science Got to do With It?**

How Ecosystems Change

Inquiry How did this happen?

This object was once part of a mining system used to move copper and iron ore. Today, so many forest plants have grown around it that it is barely recognizable. How do you think this happened? What do you think this object will look like after 500 more years?

How do communities change?

An ecosystem can change over time. Change usually happens so gradually that you might not notice differences from day to day.

1. Your teacher has given you **two pictures of ecosystem communities.** One is labeled *A* and the other is labeled *B*.

2. Imagine community A changed and became like community B. On a blank piece of **paper**, draw what you think community A might look like midway in its change to becoming like community B.

Think About This

1. What changes did you imagine? How long do you think it would take for community A to become like community B?

2. **Key Concept** Summarize the changes you think would happen as the community changed from A to B.

How Land Ecosystems Change

Have you ever seen weeds growing up through cracks in a concrete sidewalk? If they were not removed, the weeds would keep growing. The crack would widen, making room for more weeds. Over time, the sidewalk would break apart. Shrubs and vines would move in. Their leaves and branches would grow large enough to cover the concrete. Eventually, trees could start growing there.

This process is an example of **ecological succession**—*the process of one ecological community gradually changing into another.* Ecological succession occurs in a series of steps. These steps can usually be predicted. For example, small plants usually grow first. Larger plants, such as trees, usually grow last.

The final stage of ecological succession in a land ecosystem is a **climax community**—*a stable community that no longer goes through major ecological changes.* Climax communities differ depending on the type of biome they are in. In a tropical forest biome, a climax community would be a mature tropical forest. In a grassland biome, a climax community would be a mature grassland. Climax communities are usually stable over hundreds of years. As plants in a climax community die, new plants of the same species grow and take their places. The community will continue to contain the same kinds of plants as long as the climate remains the same.

Key Concept Check What is a climax community?

FOLDABLES

Fold a sheet of paper into fourths. Use two sections on one side of the paper to describe and illustrate what land might look like before secondary succession and the other side to describe and illustrate the land after secondary succession is complete.

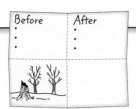

REVIEW VOCABULARY

community
all the organisms that live in one area at the same time

Primary Succession

What do you think happens to a lava-filled landscape when a volcanic eruption is over? As shown in **Figure 2,** volcanic lava eventually becomes new soil that supports plant growth. Ecological succession in new areas of land with little or no soil, such as on a lava flow, a sand dune, or exposed rock, is primary succession. *The first species that colonize new or undisturbed land are* **pioneer species.** The lichens and mosses in **Figure 2** are pioneer species.

Figure 2 Following a volcanic eruption, a landscape undergoes primary succession.

During a volcanic eruption, molten lava flows over the ground and into the water. After the eruption is over, the lava cools and hardens into bare rock.

Lichen spores carried on the wind settle on the rock. Lichens release acid that helps break down the rock and create soil. Lichens add nutrients to the soil as they die and decay.

Airborne spores from mosses and ferns settle onto the thin soil and add to the soil when they die. The soil gradually becomes thick enough to hold water. Insects and other small organisms move into the area.

After many years the soil is deep and has enough nutrients for grasses, wildflowers, shrubs, and trees. The new ecosystem provides habitats for many animals. Eventually, a climax community develops.

Secondary Succession

In areas where existing ecosystems have been disturbed or destroyed, secondary succession can occur. One example is forestland in New England that early colonists cleared hundreds of years ago. Some of the cleared land was not planted with crops. This land gradually grew back to a climax forest community of beech and maple trees, as illustrated in **Figure 3.**

 Reading Check Where does secondary succession occur?

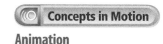 **Concepts in Motion**

Animation

Figure 3 When disturbed land grows back, secondary succession occurs.

Settlers in New England cleared many acres of forests to create cropland. In places where people stopped planting crops, the forest began to grow back.

Seeds of grasses, wildflowers, and other plants quickly began to sprout and grow. Young shrubs and trees also started growing. These plants provided habitats for insects and other small animals, such as mice.

White pines and poplars were the first trees in the area to grow to their full height. They provided shade and protection to slower growing trees, such as beech and maple.

Eventually, a climax community of beech and maple trees developed. As older trees die, new beech and maple seedlings grow and replace them.

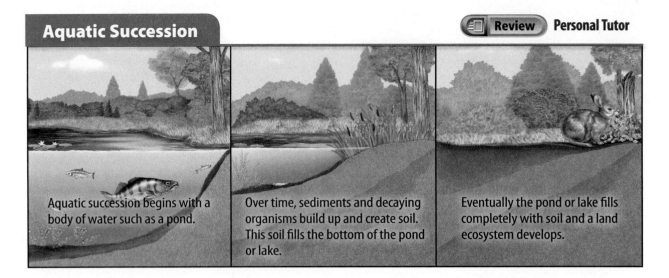

Aquatic succession begins with a body of water such as a pond.

Over time, sediments and decaying organisms build up and create soil. This soil fills the bottom of the pond or lake.

Eventually the pond or lake fills completely with soil and a land ecosystem develops.

Figure 4 The water in a pond is slowly replaced by soil. Eventually, land plants take over and the pond disappears.

How Freshwater Ecosystems Change

Like land ecosystems, freshwater ecosystems change over time in a natural, predictable process. This process is called aquatic succession.

Aquatic Succession

Aquatic succession is illustrated in **Figure 4.** Sediments carried by rainwater and streams accumulate on the bottoms of ponds, lakes, and wetlands. The decomposed remains of dead organisms add to the buildup of soil. As time passes, more and more soil accumulates. Eventually, so much soil has collected that the water disappears and the area becomes land.

 Key Concept Check What happens to a pond, a lake, or a wetland over time?

Eutrophication

As decaying organisms fall to the bottom of a pond, a lake, or a wetland, they add nutrients to the water. **Eutrophication** (yoo troh fuh KAY shun) *is the process of a body of water becoming nutrient-rich.*

Eutrophication is a natural part of aquatic succession. However, humans also contribute to eutrophication. The fertilizers that farmers use on crops and the waste from farm animals can be very high in nutrients. So can other forms of pollution. When fertilizers and pollution run off into a pond or lake, nutrient concentrations increase. High nutrient levels support large populations of algae and other microscopic organisms. These organisms use most of the dissolved oxygen in the water and less oxygen is available for fish and other pond or lake organisms. As a result, many of these organisms die. Their bodies decay and add to the buildup of soil, speeding up succession.

WORD ORIGIN

eutrophication
from Greek *eutrophos*, means "nourishing"

Lesson 3 Review

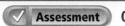
Visual Summary

Ecosystems change in predictable ways through ecological succession.

The final stage of ecological succession in a land ecosystem is a climax community.

The final stage of aquatic succession is a land ecosystem.

FOLDABLES®

Use your lesson Foldable to review the lesson. Save your Foldable for the project at the end of the chapter.

What do you think **NOW?**

You first read the statements below at the beginning of the chapter.

5. An ecosystem never changes.

6. Nothing grows in the area where a volcano has erupted.

Did you change your mind about whether you agree or disagree with the statements? Rewrite any false statements to make them true.

Use Vocabulary

1 **Define** *pioneer species* in your own words.

2 The process of one ecological community changing into another is _____.

3 **Compare and contrast** succession and eutrophication in freshwater ecosystems.

Understand Key Concepts

4 **Draw** a picture of what your school might look like in 500 years if it were abandoned.

5 Which process occurs after a forest fire?
 A. eutrophication
 B. photosynthesis
 C. primary succession
 D. secondary succession

Interpret Graphics

6 **Determine** What kind of succession— primary or secondary—might occur in the environment pictured to the right? Explain.

7 **Summarize Information** Copy the graphic organizer below and fill it with the types of succession an ecosystem can go through.

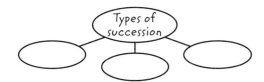

Types of succession

Critical Thinking

8 **Reflect** What kinds of abiotic factors might cause a grassland climax community to slowly become a forest?

9 **Recommend** actions people can take to help prevent the loss of wetland and estuary habitats.

Materials

paper towels

small jar

plastic wrap

jar lid

radish seeds

desk lamp

magnifying lens

Safety

A Biome for Radishes

Biomes contain plant and animal species adapted to particular climate conditions. Many organisms can live only in one type of biome. Others can survive in more than one biome. A radish is a plant grown around the world. How do you think radish seeds grow in different biomes? In this lab, you will model four different biomes and ecosystems—a temperate deciduous forest, a temperate rain forest, a desert, and a pond—and determine which biome the radishes grow best in.

Ask a Question

Which biome do radishes grow best in?

Make Observations

1. Read and complete a lab safety form.

2. Fold two pieces of paper towel lengthwise. Place the paper towels on opposite sides of the top of a small jar, as shown, with one end of each towel inside the jar and one end outside. Add water until about 10 cm of the paper towels are in the water. The area inside the jar models a pond ecosystem.

3. Place a piece of plastic wrap loosely over the end of one of the paper towels hanging over the jar's edge. Do not completely cover the paper towel. This paper towel models a temperate rain forest ecosystem. The paper towel without plastic wrap models a temperate deciduous forest.

4. Place the jar lid upside-down on the top of the jar. The area in the lid models a desert.

Form a Hypothesis

5. Observe the four biomes and ecosystems you have modeled. Based on your observations and your knowledge of the abiotic factors a plant requires, hypothesize which biome or ecosystem you think radish seeds will grow best in.

By permission of TOPS Learning Systems, www.topscience.org.

Test Your Hypothesis

6 Place three radish seeds in each biome: pond, temperate forest, temperate rain forest, and desert. Gently press the seeds to the paper towel until they stick.

7 Place your jar near a window or under a desk lamp that can be turned on during the day.

8 In your Science Journal, record your observations of the seeds and the paper towel.

9 After five days, use a magnifying lens to observe the seeds and the paper towels again. Record your observations.

Analyze and Conclude

10 **Compare and Contrast** How did the appearance of the seeds change after five days in each model biome?

11 **Critique** Evaluate your hypothesis. Did the seeds grow the way you expected? In which biome did the seeds grow the most?

12 **The Big Idea** In the biome with the most growth, what characteristics do you think made the seeds grow best?

Lab Tips

☑ Do not eat the radish seeds.

☑ If your seeds fall off the paper towel strips, do not replace them.

Communicate Your Results

Working in a group of three or four, create a table showing results for each biome. Present the table to the class.

Inquiry Extension

In this lab, you determined which biome produced the most growth of radish seeds. Seeds of different species might sprout in several different biomes. However, not all sprouted seeds grow to adulthood. Design a lab to test what conditions are necessary for radishes to grow to adulthood.

Remember to use scientific methods.

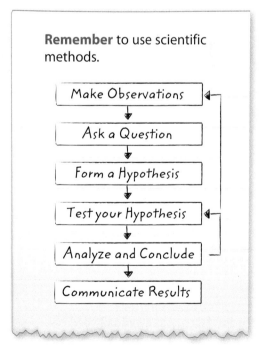

By permission of TOPS Learning Systems, www.topscience.org.

Chapter 22 Study Guide

THE BIG IDEA Each of Earth's land biomes and aquatic ecosystems is characterized by distinct environments and organisms. Biomes and ecosystems change by natural processes of ecological succession and by human activities.

Key Concepts Summary 🔑	Vocabulary
Lesson 1: Land Biomes • Each land **biome** has a distinct climate and contains animals and plants well adapted to the environment. Biomes include **deserts**, **grasslands**, tropical rain forests, **temperate** rain forests, deciduous forests, **taigas**, and **tundras**. • Humans affect land biomes through agriculture, construction, and other activities. 	**biome** p. 777 **desert** p. 778 **grassland** p. 779 **temperate** p. 781 **taiga** p. 783 **tundra** p. 783
Lesson 2: Aquatic Ecosystems  • Earth's aquatic ecosystems include freshwater and saltwater ecosystems. **Wetlands** can contain either salt water or freshwater. The **salinity** of **estuaries** varies. • Human activities such as construction and fishing can affect aquatic ecosystems.	**salinity** p. 787 **wetland** p. 790 **estuary** p. 791 **intertidal zone** p. 793 **coral reef** p. 793
Lesson 3: How Ecosystems Change • Land and aquatic ecosystems change over time in predictable processes of **ecological succession.** • Land ecosystems eventually form **climax communities.** • Freshwater ecosystems undergo **eutrophication** and eventually become land ecosystems. 	**ecological succession** p. 797 **climax community** p. 797 **pioneer species** p. 798 **eutrophication** p. 800

FOLDABLES® Chapter Project

Assemble your lesson Foldables as shown to make a Chapter Project. Use the project to review what you have learned in this chapter.

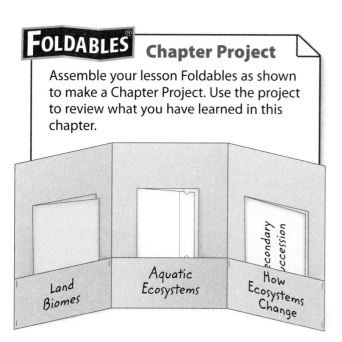

Use Vocabulary

Choose the vocabulary word that fits each description.

1. group of ecosystems with similar climate
2. area between the tropics and the polar circles
3. land biome with a layer of permafrost
4. the amount of salt dissolved in water
5. area where a river empties into an ocean
6. coastal zone between the highest high tide and the lowest low tide
7. process of one ecological community gradually changing into another
8. a stable community that no longer goes through major changes
9. the first species to grow on new or disturbed land

Link Vocabulary and Key Concepts

Concepts in Motion Interactive Concept Map

Copy this concept map, and then use vocabulary terms from the previous page and other terms from this chapter to complete the concept map.

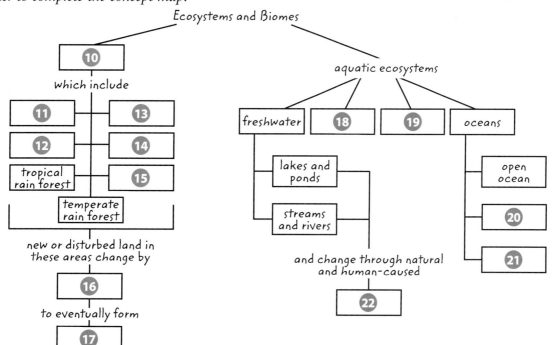

Chapter 22 Review

Understand Key Concepts 🔑

1. Where would you find plants with stems that can store large amounts of water?
   A. desert
   B. grassland
   C. taiga
   D. tundra

2. What does the pink area on the map below represent?

   A. taiga
   B. tundra
   C. temperate deciduous forest
   D. temperate rain forest

3. Where would you find trees that have no leaves during the winter?
   A. estuary
   B. tundra
   C. temperate deciduous forest
   D. temperate rain forest

4. Which biomes have rich, fertile soil?
   A. grassland and taiga
   B. grassland and tundra
   C. grassland and tropical rain forest
   D. grassland and temperate deciduous forest

5. Which is NOT a freshwater ecosystem?
   A. oceans
   B. ponds
   C. rivers
   D. streams

6. Where would you find species adapted to withstand strong wave action?
   A. estuaries
   B. wetlands
   C. intertidal zone
   D. twilight zone

7. Which ecosystem has flowing water?
   A. estuary
   B. lake
   C. stream
   D. wetland

8. Which ecosystems help protect coastal areas from flood damage?
   A. estuaries
   B. ponds
   C. rivers
   D. streams

9. Which organism below would be the first to grow in an area that has been buried in lava?

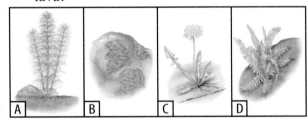

   A. A
   B. B
   C. C
   D. D

10. What is a forest called that has had the same species of trees for 200 years?
   A. climax community
   B. pioneer species
   C. primary succession
   D. secondary succession

11. What is eutrophication?
   A. decreasing nutrients
   B. decreasing salinity
   C. increasing nutrients
   D. increasing salinity

Critical Thinking

12 Compare mammals that live in tundra biomes with those that live in desert biomes. What adaptations does each group have that help them survive?

13 Analyze You are invited to go on a trip to South America. Before you leave, you read a travel guide that says the country you will be visiting has hot summers, cold winters, and many wheat farms. What biome will you be visiting? Explain your reasoning.

14 Contrast How are ecosystems in the deep water of lakes and oceans different?

15 Analyze Which type of ocean ecosystem is likely to have the highest levels of dissolved oxygen? Why?

16 Hypothesize Why are the first plants that appear in primary succession small?

17 Interpret Graphics The following climate data were recorded for a forest ecosystem. To which biome does this ecosystem likely belong?

Climate Data	June	July	August
Average temperature (°C)	16.0	16.5	17.0
Average rainfall (cm)	3.0	2.0	2.0

Writing in Science

18 Write a paragraph explaining the succession process that might occur in a small pond on a cow pasture. Include a main idea, supporting details, and concluding sentence.

REVIEW THE **BIG** IDEA

19 Earth contains a wide variety of organisms that live in different conditions. How do Earth's biomes and ecosystems differ?

20 The photo below shows Biosphere 2, built in Arizona as an artificial Earth. Imagine that you have been asked to build a biome of your choice for Biosphere 3. What biotic and abiotic features should you consider?

Math Skills ×÷

Review

— Math Practice —

Use Proportions

21 At its highest salinity, the water in Utah's Great Salt Lake contained about 14.5 g of salt in 50 g of lake water. What was the salinity of the lake?

22 The seawater in Puget Sound off the coast of Oregon has a salinity of about 24 PPT. What weight of salt is there in 1,000 g of seawater?

Record your answers on the answer sheet provided by your teacher or on a sheet of paper.

Multiple Choice

1 Which aquatic ecosystem contains a mixture of freshwater and salt water?

A coral reef

B estuary

C pond

D river

Use the diagram below to answer question 2.

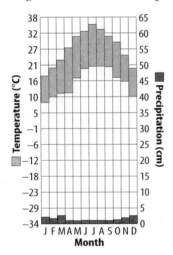

2 The diagram above most likely illustrates the climate of which biome?

A desert

B grassland

C tropical rain forest

D tundra

3 Which occurs during the first stage of ecological succession?

A eutrophication

B settlement

C development of climax community

D growth of pioneer species

4 Which biome has lost more than half its trees to logging activity?

A grassland

B taiga

C temperate deciduous forest

D tropical rain forest

Use the diagram below to answer question 5.

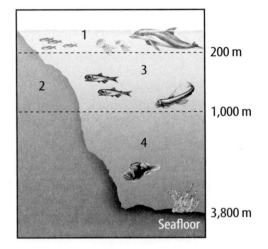

5 In the diagram above, where might you find microscopic photosynthetic organisms?

A 1

B 2

C 3

D 4

6 During aquatic succession, freshwater ponds

A become saltwater ponds.

B fill with soil.

C gain organisms.

D increase in depth.

Use the diagram below to answer question 7.

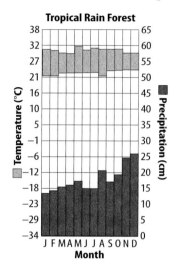

Tropical Rain Forest

7 Based on the diagram above, which is true of the tropical rain forest biome?

   A Precipitation increases as temperatures rise.

   B Rainfall is greatest mid-year.

   C Temperatures rise at year-end.

   D Temperatures vary less than rainfall amounts.

8 Which aquatic biome typically has many varieties of nesting ducks, geese, herons, and egrets?

   A coral reefs

   B intertidal zones

   C lakes

   D wetlands

Constructed Response

Use the table below to answer questions 9 and 10.

Land Biome	Climate and Plant Life	Location
Desert		
Grassland		
Taiga		
Temperate deciduous forest		
Temperate rain forest		
Tropical rain forest		
Tundra		

9 Briefly describe the characteristics of Earth's seven land biomes. List one example of each biome, including its location.

10 How does human activity affect each land biome?

Use the table below to answer question 11.

Aquatic Ecosystem	Aquatic Animal
Coastal ocean	
Coral reefs	
Estuaries	
Lakes and ponds	
Open ocean	

11 Complete the table above with the name of an aquatic animal that lives in each of Earth's aquatic ecosystems.

NEED EXTRA HELP?											
If You Missed Question...	1	2	3	4	5	6	7	8	9	10	11
Go to Lesson...	2	1	2	3	2	2	2	2	1	1	2

Using Natural Resources

THE BIG IDEA How can people protect Earth's resources?

Inquiry **A Typical Day at Work?**

These technicians are working on a wind farm off the coast of Denmark. Wind energy meets about 20 percent of Denmark's energy needs.

• How do you think wind energy works?

• What might be some advantages of using wind as an energy source?

• What might be some disadvantages?

• How can people protect Earth's resources?

Get Ready to Read

What do you think?

Before you read, decide if you agree or disagree with each of these statements. As you read this chapter, see if you change your mind about any of the statements.

1. The world's supply of coal will never run out.

2. You should include minerals in your diet.

3. Global warming causes acid rain.

4. Smog can affect human health.

5. Oil left over from frying potatoes can be used as automobile fuel.

6. Hybrid electric vehicles cannot travel far or go fast.

ConnectED Your one-stop online resource

connectED.mcgraw-hill.com

Video	WebQuest
Audio	Assessment
Review	Concepts in Motion
Inquiry	Multilingual eGlossary

Lesson 1

Reading Guide

Key Concepts 🔑
ESSENTIAL QUESTIONS

- What are natural resources?
- How do the three types of natural resources differ?

Vocabulary

natural resource p. 813

nonrenewable resource p. 814

renewable resource p. 816

inexhaustible resource p. 818

geothermal energy p. 819

 Multilingual eGlossary

 Video **BrainPOP®**

Earth's Resources

Inquiry A River in the Desert?

People in dry areas sometimes build structures such as this aqueduct to carry water to their cities. Why do you think water is such an important resource? Do you think Earth's supply of water will ever run out?

Launch Lab

15 minutes

Where does it come from?

Almost everything you use comes from natural resources, or materials that come from the environment. Can you identify the natural resources used to make a common object?

1 Read and complete a lab safety form.

2 Choose a common **object** from around your classroom or in your backpack.

3 Create a data table like the one below in your Science Journal. In the first column of your data table, list the object you will investigate.

4 In the second column, determine all the natural resources required to make the object. The example table shown here lists some natural resources used to make a pencil.

Natural Resources in a Pencil	
Object	Natural Resources Required
Pencil	1. wood
	2. graphite
	3.
	4.

Think About This

1. Which natural resource was hardest to identify? How did you figure it out?

2. Compare your data table to a classmate's. Which natural resources were on both lists?

3. **Key Concept** What type of natural resource was most common? Why might this be?

Natural Resources

You walk into a room and switch on a light. Where does the electricity to power the light come from? It might come from a power plant that burns coal or natural gas. Or, it might come from rooftop solar panels made with silicon, a mineral found in sand.

The smallest microbe and the largest whale both rely on materials and energy from the environment. The same is true for humans. People depend on the environment for food, clothing, and fuels to heat and light their homes. *Parts of the environment that supply materials useful or necessary for the survival of living things are called* **natural resources.** Natural resources include land, air, minerals, and fuels. The trees and water in **Figure 1** also are natural resources.

Key Concept Check What are natural resources?

Figure 1 All parts of the environment that are important to living things are natural resources.

Nonrenewable Resources

How often do you travel in a vehicle that runs on gasoline? Do you drink soda from aluminum cans or sip water from plastic bottles? Gasoline, aluminum, and plastic are made from nonrenewable resources.

Nonrenewable resources *are natural resources that are being used up faster than they can be replaced by natural processes.* Nonrenewable resources form slowly, usually over thousands or millions of years. If they are used faster than they form, they will run out. Nonrenewable resources include fossil fuels and minerals.

 Reading Check What characteristic makes a resource nonrenewable?

Fossil Fuels

Fossil fuels include coal, oil, and natural gas. The fossil fuels we use today formed from the decayed remains of organisms that died millions of years ago. Although fossil fuels are forming all the time, we use them much more quickly than nature replaces them. Fossil fuels form underground. As shown in **Figure 2,** coal is mined from the ground. Oil and natural gas are drilled from the ground.

Fossil fuels are used primarily as sources of energy. Many electric power plants burn coal or natural gas to heat water and make steam that powers generators. Natural gas also is used to heat homes and businesses. Gasoline, jet fuel, diesel fuel, kerosene, and other fuels are made from oil. Most plastics also are made from oil.

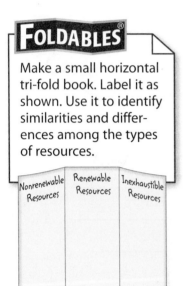

FOLDABLES®

Make a small horizontal tri-fold book. Label it as shown. Use it to identify similarities and differences among the types of resources.

| Nonrenewable Resources | Renewable Resources | Inexhaustible Resources |

Figure 2 The black rock layer is a seam of coal. It formed from the decayed remains of trees, ferns, and other swamp plants that died 300–400 million years ago.

Uranium

Concepts in Motion   Animation

Figure 3 Uranium is the fuel used to generate electricity in nuclear power plants.

✅ **Visual Check** What do you think is being emitted by the tower?

Minerals

Have you ever added fertilizer to a plant's soil? Fertilizers contain phosphorus and potassium, two minerals that promote plant growth. The human body also needs minerals for good health, including calcium and magnesium.

Minerals are nonliving substances found in Earth's crust. People use minerals for many purposes. Gypsum is used in wall board and cement. Silicon is important for the manufacture of computers and other electronic devices. Copper is used in electrical wiring.

Uranium is a mineral that can be used as a source of energy. In a nuclear power plant, such as the one shown in **Figure 3,** the nuclei of uranium atoms are split apart in a reaction known as nuclear fission. Some of the energy that held the nuclei together is released as heat, which is then used to boil water and produce steam to generate electricity.

Like fossil fuels, minerals are formed underground by geologic processes that take millions of years. For that reason, most minerals are considered nonrenewable. Some minerals, such as calcium, are plentiful. Others, such as large rubies, are rare.

✅ **Reading Check** Why are minerals nonrenewable?

Renewable Resources

Supplies of many natural resources are constantly renewed by natural cycles. The water cycle is an example. When liquid water evaporates, it rises into the atmosphere as water vapor. Water vapor condenses and falls back to the ground as rain or snow. Water is a renewable resource.

Renewable resources *are natural resources that can be replenished by natural processes at least as quickly as they are used.* These resources do not run out because they are replaced in a relatively short period of time. They include water, air, land, and living things.

Renewable resources are replenished by natural processes. Still, they must be used wisely. If people use any resource faster than it is replaced, it becomes nonrenewable. As shown in **Figure 4,** forests are sometimes nonrenewable resources.

 Reading Check In what way are renewable and nonrenewable resources similar?

Figure 4 Forests can be nonrenewable if trees are cut down faster than they can grow back.

Air

Did you know that plants produce almost all of the oxygen in the air we breathe? Oxygen is a product of photosynthesis. You might recall that photosynthesis is a series of chemical reactions in plants that use energy from light and produce sugars. Without plants, Earth's atmosphere would not contain enough oxygen to support most forms of life.

Air also contains carbon dioxide (CO_2), which plants need for photosynthesis. CO_2 is released into the air when dead plants and animals decay, when fossil fuels or wood are burned, and as a product of cellular respiration in plants and animals. Recall that cellular respiration is a series of chemical reactions that convert energy from food into a form usable by cells. Without CO_2, photosynthesis would not be possible.

Land

Fertile soil is an important resource. Topsoil is the upper layer of soil that contains most of the nutrients plants need. Gardeners know that topsoil can be replenished by the decay of plant material. The carbon, nitrogen, and other elements in the decomposing plants become available for the growth of new plants.

Topsoil can be classified as a renewable resource. However, if it is carried away by water or wind, it can take hundreds of years to rebuild.

Land resources also include wildlife and ecosystems, such as forests, grasslands, deserts, and coral reefs.

 Reading Check How is topsoil replenished by natural processes?

Water

Can you imagine a world without water? All organisms require water to live. People need a reliable supply of freshwater for drinking, washing, and irrigating crops. People also use water to run power plants and factories. Oceans, lakes, and rivers serve as major transportation routes and recreational areas. They are important habitats for many species, including some that people depend on for food.

Most of Earth's surface is covered by water. But only a small amount is freshwater that people can use, and this water must be cleaned before you can drink it. Freshwater is renewed through the water cycle, but the total amount of water on Earth always remains the same.

Has your community ever been asked to conserve water because of a drought? A drought can cause supplies of freshwater to run short. In many large cities, water is transported from hundreds of miles away to meet the needs of residents. In some parts of the world, people travel long distances every day to get the water they need.

---

inquiry MiniLab   **20 minutes**

How clean is the water?

Did you drink water or use water to wash today? Clean water is an important natural resource. It must be cleaned before it is used by people. Many water treatment plants clean water by passing it through a series of filtration steps.

1. Read and complete a lab safety form.
2. Obtain a cup of **dirty water.** Write a description of the water's appearance in a data table in your Science Journal.
3. Place a **funnel** in a **beaker.** Place a **paper towel** in the funnel. Pour the water through the paper towel. Record your observations in the table.
4. Repeat step 3, but use two layers of paper towels. Record your observations.
5. Repeat step 3, but use four layers of paper towels. Record your observations.

Step	Observations
Initial water	
1st filtration	
2nd filtration	
3rd filtration	

Analyze and Conclude

1. **Observe** How did the appearance of the water change with each filtration?

2. **Recommend** Do you think the water you produced in step 4 is clean enough to use in your home? Why or why not?

3. 🔑 **Key Concept** Summarize how people and other living things use water.

Use Percentages

Converting a ratio to a percentage often makes it easier to visualize a set of numbers. For example, in 2007, 101.5 quadrillion units (quads) of energy were used in the United States. Of that, 6.813 quads were produced from renewable energy sources. What percentage of U.S. energy was produced from renewable energy sources?

Set up a ratio of the part over the whole.

$$\frac{6.813 \text{ quads}}{101.5 \text{ quads}}$$

Rewrite the fraction as a decimal.

$$\frac{6.813 \text{ quads}}{101.5 \text{ quads}} = 0.0671$$

Multiply by 100 and add %.

$$0.0671 \times 100 = 6.71\%$$

Practice

Of the 101.5 quads of energy used in 2007, 0.341 quads were from wind energy. What percentage of U.S. energy came from wind?

 Review

- **Math Practice**
- **Personal Tutor**

Inexhaustible Resources

An **inexhaustible resource** *is a natural resource that will not run out, no matter how much of it people use.* Energy from the Sun, solar energy, is inexhaustible. So is wind, which is generated by the Sun's uneven heating of Earth's lower atmosphere. Another inexhaustible resource is thermal energy from within Earth.

Key Concept Check How do inexhaustible resources differ from renewable and nonrenewable resources?

Solar Energy

Without heat and light from the Sun, life as it is on Earth would not be possible. If you've studied food chains, you know that energy from the Sun is used by plants and other producers during photosynthesis to make food. Consumers are organisms that get energy by eating producers or other consumers. The energy in food chains always is traced back to the Sun.

Solar energy can be harnessed for many uses. Greenhouses trap heat. They make it possible to grow warm-weather plants in cool climates. Solar cookers concentrate the Sun's heat to cook food. Large solar-power plants provide electricity to many homes. Solar energy also can be used to heat water for individual homes, as shown in **Figure 5.**

Energy from an Inexhaustible Resource 🔑

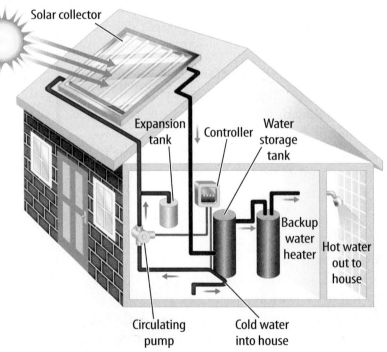

Figure 5 A solar water heater uses energy from the Sun to heat water. The hot water can be stored in a tank until it is needed.

Visual Check In which part of the system is water heated by the Sun?

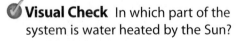

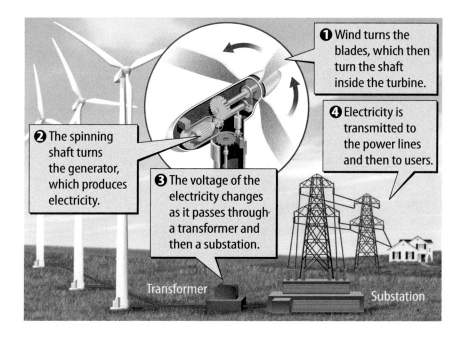

① Wind turns the blades, which then turn the shaft inside the turbine.

② The spinning shaft turns the generator, which produces electricity.

③ The voltage of the electricity changes as it passes through a transformer and then a substation.

④ Electricity is transmitted to the power lines and then to users.

Transformer

Substation

◀ **Figure 6** The wind spins the blades of a wind turbine, which in turn powers a generator that produces electricity.

✔ **Visual Check** What happens to the electricity as it passes through a transformer and a substation?

Wind Power

What do sailboats, kites, and windmills have in common? All are powered by wind—the movement of air over Earth's surface. Wind is an inexhaustible resource produced by the uneven heating of the atmosphere by the Sun.

If you live in an area with frequent, strong winds, you might have seen giant wind turbines. These turbines, such as the ones shown in **Figure 6,** can be used to produce electricity.

Geothermal Energy

Another inexhaustible resource is geothermal energy. **Geothermal energy** *is thermal energy from within Earth.* Pockets of molten rock, or magma, rise close to the surface of some parts of Earth's crust. The magma heats underground water and rocks. The heated water produces steam used to generate electricity. In California and other regions, geothermal energy produces electricity on a large scale, as shown in **Figure 7.**

WORD ORIGIN

geothermal
from Greek *geo-,* means "earth"; and Greek *therme,* means "heat"

Figure 7 Geothermal power plants use heat from within Earth to generate electricity. ▼

((◦ **Concepts in Motion** **Animation**

Lesson 1 Review

Visual Summary

Living things depend on natural resources such as water, air, and land to meet their needs.

Water is considered a renewable resource.

Wind energy can be transformed into electricity.

FOLDABLES

Use your lesson Foldable to review the lesson. Save your Foldable for the project at the end of the chapter.

What do you think NOW?

You first read the statements below at the beginning of the chapter.

1. The world's supply of coal will never run out.

2. You should include minerals in your diet.

Did you change your mind about whether you agree or disagree with the statements? Rewrite any false statements to make them true.

Use Vocabulary

1 Parts of the environment that are important to the survival of living things are _____.

2 **Define** *nonrenewable resource* in your own words.

3 **Distinguish** between renewable and inexhaustible resources.

Understand Key Concepts

4 Which is a nonrenewable resource?
   A. freshwater   C. sunlight
   B. natural gas   D. wood

5 **Explain** why inexhaustible resources also could be considered renewable.

Interpret Graphics

6 **Identify** the natural resource the device below uses, and explain how it works.

7 **Organize Information** Copy the graphic organizer below, and use it to list ways people use sunlight as a natural resource.

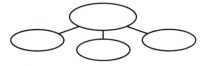

Critical Thinking

8 **Hypothesize** What measures could be taken on a farm to ensure that topsoil remains renewable?

Math Skills ×÷+   **Review**
— Math Practice —

9 Of the 101.5 quads of energy used in 2007, only 0.081 quads were from solar energy. What percentage of U.S. energy came from solar energy?

Clean Energy from Underground

Using Geothermal Energy to Heat—and Cool

AMERICAN MUSEUM OF NATURAL HISTORY

Most of the energy we use comes from burning fossil fuels such as coal and oil. This releases carbon dioxide (CO_2) into the atmosphere, which causes global surface temperatures to rise. To lower CO_2 emissions, some people are switching to clean, renewable energy sources. One source is geothermal energy, thermal energy from inside Earth.

Geothermal Heat Pumps Even as the air cools in winter and warms in summer, temperatures a few meters below Earth's surface stay pretty much the same—around 13°C. Geothermal heat pumps use the temperature difference between air and ground to heat or cool buildings, depending on the season. A geothermal heat pump moves fluid through pipes from a building into the ground and then back again. In winter, the fluid carries thermal energy from the ground to warm the building. In summer, it moves thermal energy from the building to the ground and returns with cooler fluid. Each year, approximately 50,000 geothermal heat pumps are installed in the United States.

Outside air temperature less than 13°C

Warmed air circulates through the house.

Cooler air returns to furnace.

Warmed air circulates through the house.

Cooler air returns to furnace.

Fluid releases thermal energy to the circulating air.

Cold fluid enters ground pipes.

Ground temperature around 13°C

Fluid absorbs thermal energy from the warm ground and flows back into the house.

Geothermal Power Plants Geothermal energy also can produce electricity. At one time, geothermal power plants were located only near geysers and hot springs where geothermal energy is released near Earth's surface. Today, they can be almost anywhere. These plants pump water far underground where temperatures reach up to 200°C. The water boils, and the steam is captured and brought to the surface. The steam turns turbines, which run generators that produce electricity. And it's clean energy, too: geothermal plants release only 1 percent of the carbon dioxide that coal-burning power plants produce.

It's Your Turn

RESEARCH Geothermal heat pumps have been available since the 1940s. Why do you think more homes are not using them? Research this question, and write a short report about your findings.

Pollution

Reading Guide

Key Concepts

ESSENTIAL QUESTIONS

- How does pollution affect air resources?
- How does pollution affect water resources?
- How does pollution affect land resources?

Vocabulary

pollution p. 823

ozone layer p. 824

photochemical smog p. 824

global warming p. 825

acid precipitation p. 825

  Multilingual eGlossary

Video

- BrainPOP®
- What's Science Got to do With It?

Inquiry Orange Drink?

Runoff from a mine turned the water in this stream orange. How do you think the runoff affects the organisms that live in the stream? How do you think it affects the organisms that rely on the stream as a source of freshwater?

How do air pollutants move?

Small particles of pollutants can be transported by air movement. Once a pollutant is in the air, how far can it travel?

1. Read and complete a lab safety form.
2. Use a **tape measure** to determine the distance from your desk to the **lab candle.** Record your measurement in your Science Journal.
3. As soon as your teacher blows out the candle, start a **timer.**
4. Stop the timer when you smell the blown-out candle. Record the time in your Science Journal.

Think About This

1. Divide your distance from the lab candle by the time it took you to smell the blown-out candle. How fast did the smell move?

2. Compare your results with students in different parts of the room. Why do you think the speeds varied?

3. 🔑 **Key Concept** How do you think the movement of the smell from the blown-out candle is similar to the movement of a pollutant in the air?

What is pollution?

What happens when smoke gets in the air or toxic chemicals leak into soil? Smoke is a mixture of gases and tiny particles that make breathing difficult, especially for people who have health problems. Toxic chemicals that leak into soil can kill plants and soil organisms. These substances cause pollution. **Pollution** *is the contamination of the environment with substances that are harmful to life.* An example of pollution is shown in **Figure 8.** The oil-covered animals might not survive. Other wildlife also are affected negatively, including fish that people rely on for food.

Most pollution occurs because of human actions, such as burning fossil fuels or spilling toxic materials. However, pollution also can come from natural disasters. Wildfires create smoke. Volcanic eruptions send ash and toxic gases into the atmosphere. Regardless of its source, pollution affects air, water, and land resources.

Figure 8 Oil spills pollute water and harm wildlife.

Air Pollution

Many large cities issue alerts about air quality when air pollution levels are high. On such days, people are asked to avoid activities that contribute to air pollution, such as driving cars, using gasoline-powered lawn mowers, or cooking on charcoal grills. To avoid breathing problems, people also are advised to exercise in the early morning when the air is cleaner. Air pollution that can affect human health and recreational activities can be caused by ozone loss, photochemical smog, global warming, and acid precipitation.

Ozone Loss

Ozone is a molecule composed of three oxygen atoms. In the upper atmosphere, it forms a protective layer around Earth. *The* **ozone layer** *prevents most harmful ultraviolet (UV) radiation from reaching Earth.* UV radiation from the Sun can cause cancer and cataracts and can damage crops.

In the 1980s, scientists warned that Earth's protective ozone layer was getting thinner. The problem was caused primarily by chlorofluorocarbons (CFCs). CFCs are compounds used in refrigerators, air conditioners, and aerosol sprays. Governments around the world have phased out the use of CFCs and other ozone-depleting gases. As a result, the ozone layer is expected to recover within several decades.

Photochemical Smog

Sunlight reacts with waste gases from the burning of fossil fuels and forms a type of air pollution called **photochemical smog.** As shown in **Figure 9,** smog darkens the air and also can smell bad. It is formed of particles and gases that irritate the respiratory system. One of the gases in smog is ozone. In the upper atmosphere, ozone is helpful. But in the lower atmosphere, it is a pollutant that can harm organisms and cause lung damage.

WORD ORIGIN · · · · · · · · · · ·

photochemical smog
from Greek *photo-*, means "light"; Latin *chemic*, means "alchemy"; and modern English *smog*, blend of "smoke" and "fog"

Figure 9 🔑
Photochemical smog can worsen throughout the day as chemicals continue to react with sunlight.

✓ **Visual Check** What activities contribute to the formation of smog?

Sunlight

Photochemical smog

Waste gases produced by vehicles burning fossil fuels

Global Warming

You might have heard news reports about the melting of glaciers and sea ice. Earth is getting warmer. **Global warming** *is the scientific observation that Earth's average surface temperature is increasing.* Global warming can lead to climate change—changing weather conditions, changes to ecosystems and food webs, increases in the number and severity of floods and droughts, and increased coastal flooding as sea ice melts and sea level rises.

Data indicate that Earth's average surface temperature and increases in atmospheric carbon dioxide (CO_2) follow the same general trend. CO_2 is a greenhouse gas. This means it traps heat, helping to keep Earth warm. Greenhouse gases occur naturally. Without them, Earth would be too cold to support life. But human activities add greenhouse gases to the atmosphere, especially CO_2 from the burning of fossil fuels. Most scientists, including those on the United Nations Intergovernmental Panel on Climate Change, agree that increases in atmospheric CO_2 are contributing to global warming.

Acid Precipitation

Gases produced by the burning of fossil fuels also create other forms of air pollution, including acid precipitation. **Acid precipitation** *is acidic rain or snow that forms when waste gases from automobiles and power plants combine with moisture in the air.* Coal-burning power plants produce sulfur dioxide gas that combines with moisture to form sulfuric acid. Cars and trucks produce nitrous oxide gases that form nitric acid. Acid precipitation pollutes soil and can kill plants, including trees, as shown in **Figure 10**. It also contributes to water pollution and can damage buildings.

 Key Concept Check How does pollution affect air resources?

Figure 10 Acid rain can harm soil organisms and plant roots.

Visual Check How did acid rain affect this ecosystem?

Review Personal Tutor

MiniLab

20 minutes

How fast can you turn a sand castle into sediment?

Runoff can move sediment into streams. Sediment blocks stream flow, clogs the feeding structures of animals, and decreases the amount of light for aquatic plants. How does the flow of water affect rates of sedimentation?

1. Read and complete a lab safety form.

2. Use a **foam cup** to build a **sand** castle in a **plastic container.** Measure its height with a **metric ruler.** Record the data in your Science Journal.

3. Fill a **spray bottle** with water. Adjust the tip of the bottle so it sprays a mist.

4. Using a **timer,** spray your sand castle for 30 s. Measure and record the height of your sand castle.

5. Readjust the tip of the spray bottle so it sprays a stream of water. Then, rebuild your castle with fresh sand and repeat step 4.

6. Rebuild your sand castle with fresh sand. Poke three holes in the bottom of the foam cup with a **pencil.** Put your finger over the holes and fill the cup with water. Repeat step 4, letting water run out of the holes onto your castle.

Analyze and Conclude

1. **Evaluate** Which trial caused the largest change in the height of the sand castle?

2. **Model** What natural events could each of your trials represent?

3. **Key Concept** How might these natural events affect the quality of water resources?

Water Pollution

Have you ever seen a stream covered with thick green algae? The stream might have been polluted with fertilizers from nearby lawns or farms. It might contain chemicals from nearby factories. Water pollution can come from chemical runoff and other agricultural, residential, and industrial sources.

Wastewater

You might have been warned not to pour paint or used motor oil into storm drains. In most cities, rainwater that flows into storm drains goes directly into nearby waterways. Materials that go in the drain, including grease and oil washed from the street, can contribute to water pollution.

The wastewater that drains from showers, sinks, and toilets contains harmful viruses and bacteria. To safeguard health, this wastewater usually is purified in a sewage treatment plant before it is released into streams or used to irrigate crops. In some parts of the world, there is little or no sewage treatment. People might have to use polluted water.

Wastewater that comes from industries and mining operations also contains pollutants. It requires treatment before it can be returned to the environment. Even after treatment, some harmful substances might remain and impact water quality.

Runoff and Sediments

When it rains, water can flow over the land. This water, called runoff, flows across lawns and farmland. Along the way, it picks up pesticides, herbicides, and fertilizers. Runoff carries these pollutants into streams, where they can harm insects, fish, and other organisms. Runoff also carries sediment particles into streams. Too much sediment can damage stream habitats, clog waterways, and cause flooding.

Key Concept Check How does pollution affect water resources?

Figure 11 Fertilizers and irrigation water contain salts that can build up in soil, as shown in the photo on the left. The photo on the right shows a mining technique that disturbs ecosystems.

Land Pollution

Have you ever helped clean up litter? Foam containers, plastic bags, bottles, cans, and even furniture and appliances get dumped along roadsides. Litter is more than an eyesore. It can pollute soil and water and disturb wildlife. Sources of land pollution include homes, farms, industry, and mines.

Agriculture

Farmers use pesticides and other agricultural chemicals to help plants grow. But these chemicals become pollutants if they are used in excess or disposed of improperly. Herbicides kill weeds. But if they flow into streams, they can kill algae and plants, and harm fish and amphibians. Some farming practices contaminate soil, as shown in **Figure 11.**

Industry and Mining

Many industrial facilities, including oil refineries and ore processors, produce toxic wastes. For example, coal ash sludge is produced when coal is burned in power plants. The sludge contains mercury, lead, arsenic, and other potentially harmful metals. If toxic wastes such as these are incorrectly stored or disposed of, they contaminate soil and water. The health of people, plants, and wildlife can be affected.

Mining of fossil fuels and minerals can disturb or destroy entire ecosystems, as shown in **Figure 11.** Some coal-mining techniques can release toxic substances that were buried in rock. After the coal has been removed, the area can be restored. But it is difficult or impossible to replace the original ecosystem.

Key Concept Check How does pollution affect land resources?

Lesson 2 Review

Visual Summary

Pollution, the introduction of harmful substances into the environment, can harm humans and other living things.

Smog, ozone loss, global warming, and acid precipitation are caused by air pollutants.

Land and water can be polluted by littering and chemical runoff from homes, factories, mines, and farms.

FOLDABLES

Use your lesson Foldable to review the lesson. Save your Foldable for the project at the end of the chapter.

What do you think NOW?

You first read the statements below at the beginning of the chapter.

3. Global warming causes acid rain.

4. Smog can affect human health.

Did you change your mind about whether you agree or disagree with the statements? Rewrite any false statements to make them true.

Use Vocabulary

1. **Define** *photochemical smog* in your own words.

2. **Distinguish** between global warming and acid precipitation.

3. The layer of atmosphere that prevents UV radiation from reaching Earth is the _____.

Understand Key Concepts

4. Which is NOT a source of pollution?
   A. burning coal   C. photosynthesis
   B. mining minerals D. volcanic eruptions

5. **Explain** the difference between ozone in the lower atmosphere and ozone in the upper atmosphere.

6. **Describe** how pollution affects water resources.

Interpret Graphics

7. **Compare and Contrast** Copy and fill in the table below to compare and contrast air, water, and land pollution.

Type of Pollution	Similarities	Differences
Air		
Water		
Land		

Critical Thinking

8. **Hypothesize** The water in a stream that flows through farmland has always been clear. After a hard rain, the water in the stream became muddy. What caused the change?

9. **Apply** How do trees and other plants help lessen global warming? How might deforestation, or cutting down trees, contribute to global warming?

How can you communicate about pollution?

You have read about different types of pollutants in this chapter. Now it's your turn to communicate what you have learned. A public service announcement (PSA) is like a commercial that explains an important issue.

Materials

stopwatch

computer

Learn It

Communication of ideas is an important part of the work of scientists. A scientific idea that is not reported will not advance scientific knowledge or the public's understanding of science. Scientists often **communicate** their ideas in presentations.

Try It

1. Read and complete a lab safety form.

2. Choose a pollutant you read about in this chapter or a different pollutant in which you have an interest.

3. Research your pollutant. Find out as much as you can about how it is produced, how it enters the environment, what problems it causes, and how its effects can be reduced.

4. Write a 1-min script for a PSA that communicates the information you gathered in step 3.

5. Practice your script until you feel comfortable speaking it before a group. If recording equipment is available, record your PSA.

Apply It

6. Present your PSA to your class.

7. Take questions from the class. Ask your classmates what they learned. Record their comments in your Science Journal.

8. **Critique** your PSA. Did the class understand the message you were trying to communicate? How could you improve your presentation?

9. **Key Concept** How does the pollutant you researched affect natural resources?

Protecting Earth

Inquiry A Better View?

This image taken from a satellite shows an eruption on a volcanic island. What parts of the environment do you see in the image? How can images taken from satellites help scientists study Earth?

How can you turn trash into art?

Reusing materials helps reduce the natural resources needed to make something new. It also reduces the amount of trash discarded in landfills. Some environmentally friendly artists reuse materials to create new art. In this lab, you will create something new from objects you usually might throw away.

1. Read and complete a lab safety form.
2. Carefully consider the materials in the **trash collection.**
3. Using **craft materials,** create a piece of art out of the trash. Try to convey a message with your art. For example, your message might be "Protect Earth."
4. Display your artwork to the class.

Think About This

1. Describe your artwork. What kind of trash did you use? What kind of art did you create?
2. What message did you try to convey?
3. **Key Concept** How do you think reusing materials helps conserve resources?

Monitoring Human Impact on Earth

Earth's human population is expected to grow to 7 billion by 2012. As the population increases, so does humans' impact on the planet. Scientists, governments, and concerned citizens around the world are working to identify environmental problems, educate the public about them, and help find solutions.

Scientists collect data on a variety of environmental conditions by placing detectors on satellites, aircraft, high-altitude balloons, and ground-based monitoring stations. For example, the United States and the European Union have launched satellites into orbit around Earth to gather data on greenhouse gases, ozone, ecosystem changes, melting glaciers and sea ice, climate patterns, and ocean health.

The U.S. Environmental Protection Agency (EPA) is a government organization that monitors the health of the environment and looks for ways to reduce human impacts. The EPA enforces environmental laws and supports research at universities and national laboratories. It also works with citizens and organizations to identify superfund sites—abandoned areas that have been contaminated by toxic wastes—and develops plans for cleaning them up.

Key Concept Check How can people monitor resource use?

FOLDABLES

Make a small shutter-fold book. Label it as shown. Use it to identify technology and methods that protect natural resources.

Monitoring Natural Resources

Conserving Natural Resources

Developing Technologies

Many technologies have been developed to protect Earth's resources, and more are on the way. These advances often focus on saving energy and reducing pollution.

Water-Saving Technologies

It takes energy to clean water and to transport it to homes and businesses. So technologies that conserve water also save energy. Low-flow showerheads and toilets help reduce water use. Drip irrigation systems, such as the one shown in **Figure 12,** decrease water waste.

Energy-Saving Technologies

Saving energy can make Earth's supply of fossil fuels last longer. Relying more on renewable energy sources can reduce fossil fuel use. Some of these sources can be expensive, but designs are constantly improving and costs are going down. Researchers estimate that solar electricity soon will cost no more than electricity produced by burning fossil fuels. Burning fewer fossil fuels also creates less pollution.

Other energy-saving advances include compact fluorescent lightbulbs (CFLs). They use about one-fourth the energy of incandescent bulbs and can last ten times longer. In 2007, Americans reduced greenhouse gas emissions by an amount equal to removing 2 million cars from the road just by switching to CFLs.

Figure 12 Drip irrigation slowly delivers water to the roots of plants. Less water is lost to runoff and evaporation.

Inquiry MiniLab

20 minutes each day

What's in the air?

Air pollution made of particles that float in the air is called particulate matter, or PM. The Clean Air Act requires the EPA to monitor PM.

1. Read and complete a lab safety form.

2. To make a PM collector, coat two **plastic container lids** with a layer of **petroleum jelly.**

3. Leave each PM collector in a different location around your school. Record the location, the date, and the time in your Science Journal.

4. On the following day, retrieve the collectors. Record the date and time.

5. Use a **magnifying lens** to observe the PM. Record your observations.

Analyze and Conclude

1. **Describe** What types of PM did you find on your collectors?

2. **Compare** the amount and type of PM found in the different locations. Formulate a reason for any differences you observe.

3. **Key Concept** What conclusions about air quality in your school can you draw from your data?

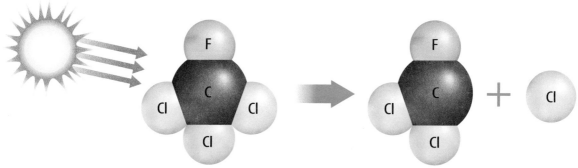

Sunlight reacts with a CFC molecule, causing a chlorine atom to break away.

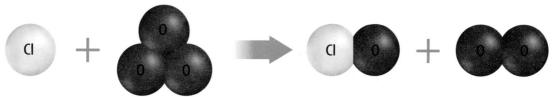

The chlorine atom reacts with and breaks apart an ozone molecule.

CFC Replacements

You have read that CFCs cause thinning of the ozone layer. How does this happen? The chlorine atoms in CFC molecules react with sunlight to destroy ozone, as shown in **Figure 13.** All CFCs soon will be phased out and replaced with chemicals that do not contain chlorine. Replacements include hydrofluorocarbons (HFCs) and perfluorocarbons (PFCs). Even after CFCs are no longer in use, it will take decades for the ozone layer to recover.

 Reading Check What steps have been taken to reverse the thinning of the ozone layer?

Alternative Fuels

Gasohol and biodiesel are alternative fuels that help reduce humans' use of fossil fuels. They also help reduce air pollution.

Gasohol is a mixture of 90 percent gasoline and 10 percent ethanol. Ethanol is alcohol made from corn, sugar cane, or other plants. Using gasohol helps reduce emissions of carbon monoxide, an air pollutant that contributes to smog. The carbon in ethanol comes from plants rather than fossil fuels. So, using gasohol can help reduce emissions that contribute to global warming.

Biodiesel is made from renewable resources, primarily vegetable oils and animal fats—including oil left over from frying foods in restaurants. Biodiesel can be burned in diesel engines in farm and industrial machinery, trucks, and cars. It produces fewer pollutants than regular diesel fuel, and it reduces CO_2 emissions by 78 percent.

Figure 13 When CFC molecules reach the upper atmosphere, sunlight breaks off chlorine atoms. Each free chlorine atom can destroy an ozone molecule and prevent another from forming.

Visual Check How do CFCs affect ozone molecules?

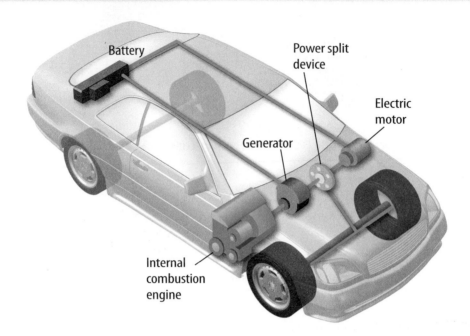

Battery

Power split device

Electric motor

Generator

Internal combustion engine

Figure 14 A hybrid vehicle uses a battery to power an electric motor. A small gasoline engine provides additional power.

Visual Check What are the power sources in a hybrid vehicle?

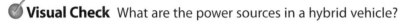

SCIENCE USE V. COMMON USE

hybrid

Science Use an offspring of two animals or plants of different breeds or species

Common Use something that has two different components performing essentially the same task

Automobile Technologies

If you were buying a car, you would want to know how many miles it travels per gallon of fuel—miles per gallon, or mpg. The higher a car's mpg, the less pollution it will add to the environment. A car with a high mpg also will use up fewer fossil-fuel resources.

One of your choices might be a **hybrid** electric vehicle (HEV). HEVs combine a small gasoline engine with an electric motor powered by batteries, as shown in **Figure 14.** HEVs run on battery power as much as possible, with a boost from the engine for longer trips, higher speeds, and steep hills. The engine also charges the batteries. HEVs can get up to twice the mileage of a conventional car—close to 50 mpg in some recent car models.

In the future, another automobile alternative might be a fuel-cell vehicle (FCV). Inside a fuel cell, oxygen from the air chemically combines with hydrogen to produce electricity. The primary waste product is water. Tailpipe emissions from FCVs are nearly pollution-free. However, obtaining hydrogen fuel requires using methane or other fossil fuels. Researchers are looking for alternatives.

Reading Check Compare HEVs and FCVs.

Making a Difference

Do you turn off the lights when you leave a room or recycle bottles and cans? If so, you are helping reduce your impact on the environment. You can help protect Earth's resources in other ways as well, such as cleaning up a stream, educating others about environmental issues, analyzing the choices you make as a consumer, and following some of the suggestions you will read about next.

Sustainability

When people talk about environmental issues, they often use the word *sustainability*. **Sustainability** *means meeting human needs in ways that ensure future generations also will be able to meet their needs.* When you turn off the lights as you leave a room, you are saving energy—and you are also helping to ensure a sustainable future. **Figure 15** shows other actions that lead toward a sustainable future.

 Reading Check What is sustainability?

Sustainable Actions

Figure 15 Planting trees, composting, and picking up litter help sustain the environment.

Restore and Rethink

Restoring damaged habitats and ecosystems to their original state is one way to make a difference. For example, picking up trash can restore water habitats.

You also can rethink the way you perform everyday activities. Instead of riding in a vehicle to nearby places, you could ride your bike or walk.

Reduce and Reuse

You can reduce the amount of waste you create simply by reducing the amount of material you use. For example, avoid products with too much packaging. Or, bring your own bags for carrying purchases, as shown in **Figure 16.**

Reusing items also helps reduce waste. Instead of buying new, reuse something that will work just as well. You also can donate used items to charities or resell them.

Figure 16  Reusable bags help save energy and reduce waste.

Recycle

If an item cannot be reused, you might be able to recycle it. **Recycling** is *manufacturing new products out of used products.* This process reduces wastes and extends our supply of natural resources. Computers and other electronics contain valuable metals that can be recycled, as well as toxic materials that can contribute to pollution. So recycling also helps makes sure that toxins are properly disposed of.

WORD ORIGIN ············

recycle
from Latin *re-*, means "again"; and Greek *kyklos*, means "circle"

Compost Leaves, grass clippings, and vegetable scraps can be recycled by composting. In a compost pile, these materials decay into nutrient-rich soil that goes back into the garden.

Buy Recycled Separating recyclables from the rest of the trash is just one step. To keep the cycle going, buy and use recycled products. You can find shoes, clothing, paper, and carpets made from recycled materials.

Key Concept Check How can you conserve resources?

Lesson 3 Review

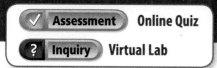
Visual Summary

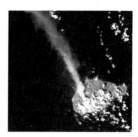

Scientists use a variety of techniques to monitor the use of natural resources, including satellites, aircraft, high-altitude balloons, and ground-based monitoring stations.

New technologies such as HEVs and alternative fuels conserve resources and produce less pollution.

People can help protect resources by reducing their use of resources, reusing products, and recycling products.

FOLDABLES®

Use your lesson Foldable to review the lesson. Save your Foldable for the project at the end of the chapter.

What do you think NOW?

You first read the statements below at the beginning of the chapter.

5. Oil left over from frying potatoes can be used as automobile fuel.

6. Hybrid electric vehicles cannot travel far or go fast.

Did you change your mind about whether you agree or disagree with the statements? Rewrite any false statements to make them true.

Use Vocabulary

1 **Define** *sustainability* in your own words.

Understand Key Concepts

2 Which produces water as its primary waste product?
   A. biodiesel   C. gasohol
   B. FCV   D. HEV

3 **Analyze** Compare the tailpipe emissions of an HEV with a car that has only a gasoline engine.

4 **Apply** What water-saving and energy-saving techniques could you use in your kitchen?

Interpret Graphics

5 **Explain** how the process below affects the environment.

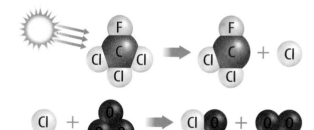

6 **Identify** Copy and fill in the graphic organizer below to identify three ways people can limit waste production.

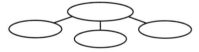

Critical Thinking

7 **Hypothesize** The same delivery truck passes you almost every morning as you walk to school. Each time it goes by, you smell fried potatoes. What could be the reason?

8 **Recommend** The school cafeteria throws out mounds of vegetable peelings and leftover food every day. The school gardener would like to plant vegetables, but the soil is too thin. What sustainable actions could you recommend?

How can you conserve a natural resource?

You have read about natural resources and about how they can be affected by human activities. Even though you are just one person, your actions can help conserve natural resources. Your task is to develop a realistic plan to conserve a natural resource.

Ask Questions

How do your daily activities affect natural resources? What actions could you take to conserve a natural resource?

Make Observations

1. Read and complete a lab safety form.

2. Select the natural resource you want to help conserve. In your Science Journal, describe how the resource is used. Also discuss the problems that threaten the natural resource.

3. Review the list of conservation activities below. Choose activities from the list or come up with your own ideas. Explain how the activities would conserve the resource you have chosen.

4. Write a plan that details how you will enact these activities in your everyday life. Describe the materials you will need and the steps you will follow. Make a time line showing how you will implement the plan.

5. Remember, your plan must be realistic. Conduct research as you do this investigation to help you learn more about natural resources and conservation plans.

Conservation Activities
Change showerheads and faucets to low-flow.
Install rain barrels to use for watering gardens.
Start a compost pile for food scraps or yard waste.
Provide recycling bins at sporting events.
Replace incandescent lights with CFLs or LEDs.
Clean up a local road or pond.
Replace car trips with bicycling or walking whenever possible.
Lower the heating thermostat in winter and raise the air-conditioner thermostat in summer.

Form a Hypothesis

6 State the major goal or goals of your plan in the form of a hypothesis.

Test your Hypothesis

7 Discuss your plan with a classmate. Does he or she think it will work? Is the plan realistic, or will it require large amounts of money, time, and resources? Modify your plan based on your classmate's input.

8 Implement your plan over a scheduled period of time. Follow your time line. Record your observations and any quantifiable data.

Analyze and Conclude

9 **Assess** How did your plan affect the natural resource? Which impacts were you able to quantify? Which impacts were difficult to quantify?

10 **Evaluate** Did you see any limitations to your plan? For example, can it be implemented on a larger scale?

11 **The Big Idea** Why is it important to actively work to conserve Earth's resources?

Lab Tips

☑ If you are having trouble deciding on a topic, think about your activities as you go about your day. What things do you do that use a lot of natural resources?

☑ Be flexible. Propose more than one way to solve a problem.

☑ Be creative, but keep your plan realistic.

Communicate Your Results

Create an assessment report of your plan. Make your report engaging and informative. Provide a comprehensive description of your results. Include a proposal for extending the time period of the plan.

Inquiry Extension

Create a 1-min PSA describing your plan. The PSA should encourage others to follow the plan. Include a graph or other visual that shows the positive impact of your plan on the natural resource. Project how this impact would increase if everyone in the class followed your plan. Project the impact of your plan on county, state, and national levels.

Remember to use scientific methods.

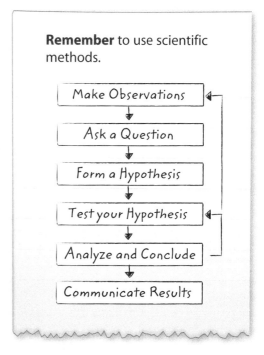

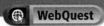

THE BIG IDEA People can protect Earth's resources by understanding how their use of natural resources affects the environment, knowing which natural resources are in limited supply, and making decisions toward a more sustainable future.

Key Concepts Summary 🔑

Lesson 1: Earth's Resources

- **Natural resources** are raw materials and forms of energy that are important to living things.

- Resources can be **renewable** or **nonrenewable.** Some renewable resources are **inexhaustible.**

Vocabulary

natural resource p. 813

nonrenewable resource p. 814

renewable resource p. 816

inexhaustible resource p. 818

geothermal energy p. 819

Lesson 2: Pollution

- Air pollutants cause **photochemical smog,** ozone loss, **global warming,** and **acid precipitation.**

- Chemical runoff can damage lakes, streams, and water supplies. Sediment runoff from land can disturb aquatic habitats.

- Litter and pollutants can contaminate soil, harm organisms, and reduce land's ability to support life. Mining can disturb ecosystems and create toxic wastes.

pollution p. 823

ozone layer p. 824

photochemical smog p. 824

global warming p. 825

acid precipitation p. 825

Lesson 3: Protecting Earth

- Satellites, aircraft, and ground-based monitoring stations collect data on pollution. The EPA monitors pollution and helps develop clean-up plans.

- People can protect Earth's resources by reducing, reusing, and **recycling.**

sustainability p. 835

recycling p. 836

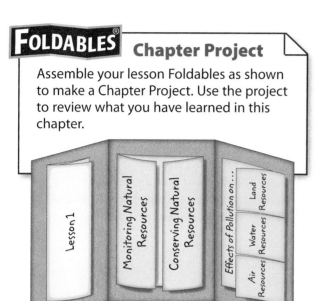

FOLDABLES® Chapter Project

Assemble your lesson Foldables as shown to make a Chapter Project. Use the project to review what you have learned in this chapter.

Use Vocabulary

1 Sunshine, oil, coal, uranium, trees, oxygen, and streams are examples of _____.

2 Distinguish between recycling, reusing, and reducing.

3 A tropical forest that takes 1,000 years to recover from being burned down is a(n) _____ resource.

4 Use the term *sustainability* in a sentence.

5 Rainfall that keeps a pond full is a(n) _____ resource.

6 Define *global warming* in your own words.

Link Vocabulary and Key Concepts

 Concepts in Motion **Interactive Concept Map**

Copy this concept map, and then use vocabulary terms from the previous page and other terms from the chapter to complete the concept map.

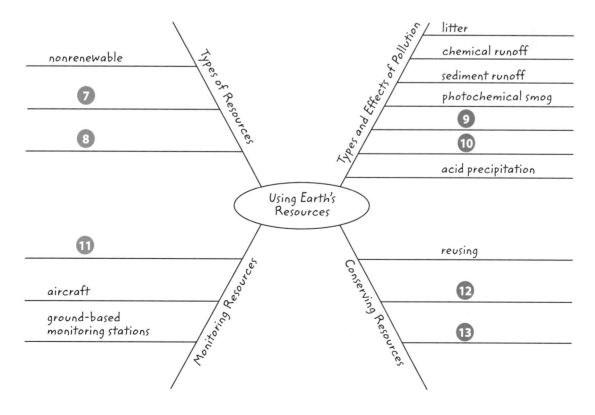

Chapter 23 Review

Understand Key Concepts

1. Which is an inexhaustible resource?
   A. air
   B. land
   C. water
   D. wind

2. What are the items below made of?

   A. inexhaustible resources
   B. nonrecyclable resources
   C. nonrenewable resources
   D. renewable resources

3. Greenhouse gases contribute to which environmental problem?
   A. acid rain
   B. global warming
   C. ozone depletion
   D. photochemical smog

4. Biodiesel can be made from which resource?
   A. ethanol
   B. gasoline
   C. hydrogen gas
   D. vegetable oil

5. Which technology produces electricity using sunlight?
   A. fuel cells
   B. solar cookers
   C. solar power plants
   D. solar water heaters

6. What kind of pollution results when sunlight reacts with sulfur dioxide produced by the burning of fossil fuels?
   A. acid rain
   B. global warming
   C. ozone depletion
   D. photochemical smog

7. Which is a sustainable action?
   A. carpooling to a game
   B. riding in a car to school
   C. running water when brushing teeth
   D. throwing out a slightly used shirt

8. What process is illustrated in the diagram below?

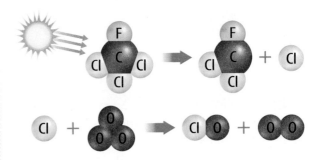

   A. acid rain
   B. global warming
   C. ozone depletion
   D. smog formation

9. Uranium is classified as what?
   A. a fossil fuel
   B. a greenhouse gas
   C. a mineral resource
   D. a renewable resource

10. Which is a renewable resource?
    A. coal
    B. sunlight
    C. water
    D. wind

Critical Thinking

11 **Classify** each of the following as renewable, nonrenewable, or inexhaustible: notebook paper, flashlight batteries, and heat from a volcano.

12 **Explain** why geothermal energy is considered an inexhaustible resource.

13 **Design** an experiment to determine how acid rain affects plants.

14 **Interpret Scientific Illustrations** Describe how the process below affects human health.

15 **Create** a poster explaining why items such as used motor oil or leftover paint should be recycled rather than poured down a storm drain or dumped on soil.

16 **Summarize** how people can help prevent land pollution.

17 **Compare** the emissions of a gasoline-powered automobile, a hybrid electric vehicle, and a fuel-cell vehicle. Which vehicle contributes the least amount of air pollution?

18 **Give** an example of how people can reduce the amount of trash they produce.

Writing in Science

19 **Write** a letter to a younger student about sustainability. Explain what it means, why it is important to his or her generation, and how it involves considering future needs, not just current ones.

REVIEW THE BIG IDEA

20 Give examples of what a government and an individual could do to protect Earth's resources.

21 How can using wind energy help conserve Earth's resources?

Math Skills

 Review

— **Math Practice** —

Use Percentages

22 Between 2003 and 2006, the amount of U.S. energy produced from renewable sources increased from 6.15 quads to 6.91 quads. What was the percentage increase?

23 Wind energy usage increased from 0.115 quads to 0.341 quads between 2003 and 2007. What was the percentage increase?

24 In 2006, only 6.92 percent of U.S. energy was produced from renewable sources. If the total energy consumption was 99.8 quads, how much energy was produced from renewable sources?

Record your answers on the answer sheet provided by your teacher or on a sheet of paper.

Multiple Choice

1 Why is coal a nonrenewable resource?

   A Coal is used faster than it forms.

   B Coal cannot be recycled like glass or plastic.

   C Humans do not know how to make coal.

   D Coal formed in the past and no longer forms today.

2 Which is an effect of photochemical smog?

   A global warming

   B lung damage

   C skin cancer

   D acid precipitation

Use the diagram to answer questions 3 and 4.

3 The diagram shows a chlorofluorocarbon (CFCs) reaction. Where does this reaction occur?

   A in water

   B on land

   C in an automobile

   D in the atmosphere

4 How does this reaction affect the environment?

   A It depletes the ozone layer.

   B It pollutes water.

   C It produces acid rain.

   D It produces photochemical smog.

5 Which government agency monitors the health of the environment and works to reduce human impacts?

   A CFC

   B CFL

   C EPA

   D HEV

6 Which material is NOT a fossil fuel?

   A coal

   B copper

   C natural gas

   D oil

Use the diagram to answer question 7.

7 Nika is reading a pamphlet about the proper disposal of different materials. The pamphlet includes the image shown above. What would happen if someone performed the action shown in the image?

   A The oil would cause photochemical smog.

   B The oil would be washed into nearby streams or lakes.

   C The oil would evaporate and cause ozone depletion.

   D The oil would remain in the storm drain and not cause harm.

8 Which action recycles yard waste and kitchen waste?

  A composting

  B planting trees

  C using drip irrigation

  D using compact fluorescent lightbulbs

Use the diagram to answer question 9.

9 What type of energy resource is shown in the diagram?

  A geothermal

  B solar

  C water

  D wind

10 Which is a renewable resource that can become nonrenewable if it is used up too quickly?

  A oil

  B forests

  C coal

  D natural gas

Constructed Response

11 What action could you take to help conserve resources and reduce the amount of waste that enters landfills?

Use the diagram to answer question 12.

12 Which part of the environment is most affected by the activity shown here? Explain your answer.

13 Describe how rethinking your everyday activities contributes to a sustainable future.

14 Identify one renewable resource and one inexhaustible resource. Then, describe similarities and differences between the two resources.

NEED EXTRA HELP?														
If You Missed Question...	1	2	3	4	5	6	7	8	9	10	11	12	13	14
Go to Lesson...	1	2	3	3	3	1	3	3	1,3	1	3	2,3	3	1

Student Resources

For Students and Parents/Guardians

These resources are designed to help you achieve success in science. You will find useful information on laboratory safety, math skills, and science skills. In addition, science reference materials are found in the Reference Handbook. You'll find the information you need to learn and sharpen your skills in these resources.

Table of Contents

Scientific Methods

SCIENCE SKILL HANDBOOK

MATH SKILL HANDBOOK

FOLDABLES HANDBOOK

REFERENCE HANDBOOK

GLOSSARY/ GLOSARIO

INDEX

Scientists use an orderly approach called the scientific method to solve problems. This includes organizing and recording data so others can understand them. Scientists use many variations in this method when they solve problems.

Identify a Question

The first step in a scientific investigation or experiment is to identify a question to be answered or a problem to be solved. For example, you might ask which gasoline is the most efficient.

Gather and Organize Information

After you have identified your question, begin gathering and organizing information. There are many ways to gather information, such as researching in a library, interviewing those knowledgeable about the subject, and testing and working in the laboratory and field. Fieldwork is investigations and observations done outside of a laboratory.

Researching Information Before moving in a new direction, it is important to gather the information that already is known about the subject. Start by asking yourself questions to determine exactly what you need to know. Then you will look for the information in various reference sources, like the student is doing in **Figure 1.** Some sources may include textbooks, encyclopedias, government documents, professional journals, science magazines, and the Internet. Always list the sources of your information.

Figure 1 The Internet can be a valuable research tool.

Evaluate Sources of Information Not all sources of information are reliable. You should evaluate all of your sources of information, and use only those you know to be dependable. For example, if you are researching ways to make homes more energy efficient, a site written by the U.S. Department of Energy would be more reliable than a site written by a company that is trying to sell a new type of weatherproofing material. Also, remember that research always is changing. Consult the most current resources available to you. For example, a 1985 resource about saving energy would not reflect the most recent findings.

Sometimes scientists use data that they did not collect themselves, or conclusions drawn by other researchers. This data must be evaluated carefully. Ask questions about how the data were obtained, if the investigation was carried out properly, and if it has been duplicated exactly with the same results. Would you reach the same conclusion from the data? Only when you have confidence in the data can you believe it is true and feel comfortable using it.

Interpret Scientific Illustrations As you research a topic in science, you will see drawings, diagrams, and photographs to help you understand what you read. Some illustrations are included to help you understand an idea that you can't see easily by yourself, like the tiny particles in an atom in **Figure 2.** A drawing helps many people to remember details more easily and provides examples that clarify difficult concepts or give additional information about the topic you are studying. Most illustrations have labels or a caption to identify or to provide more information.

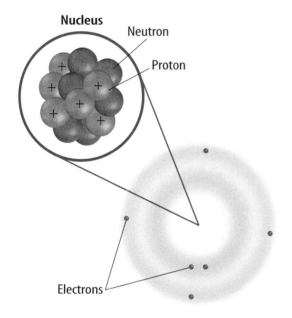

Figure 2 This drawing shows an atom of carbon with its six protons, six neutrons, and six electrons.

Concept Maps One way to organize data is to draw a diagram that shows relationships among ideas (or concepts). A concept map can help make the meanings of ideas and terms more clear, and help you understand and remember what you are studying. Concept maps are useful for breaking large concepts down into smaller parts, making learning easier.

Network Tree A type of concept map that not only shows a relationship, but how the concepts are related is a network tree, shown in **Figure 3.** In a network tree, the words are written in the ovals, while the description of the type of relationship is written across the connecting lines.

When constructing a network tree, write down the topic and all major topics on separate pieces of paper or notecards. Then arrange them in order from general to specific. Branch the related concepts from the major concept and describe the relationship on the connecting line. Continue to more specific concepts until finished.

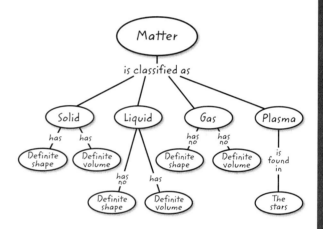

Figure 3 A network tree shows how concepts or objects are related.

Events Chain Another type of concept map is an events chain. Sometimes called a flow chart, it models the order or sequence of items. An events chain can be used to describe a sequence of events, the steps in a procedure, or the stages of a process.

When making an events chain, first find the one event that starts the chain. This event is called the initiating event. Then, find the next event and continue until the outcome is reached, as shown in **Figure 4** on the next page.

SCIENCE SKILL HANDBOOK

MATH SKILL HANDBOOK

FOLDABLES HANDBOOK

REFERENCE HANDBOOK

GLOSSARY/ GLOSARIO

INDEX

SCIENCE SKILL HANDBOOK

MATH SKILL HANDBOOK

FOLDABLES HANDBOOK

REFERENCE HANDBOOK

GLOSSARY/ GLOSARIO

INDEX

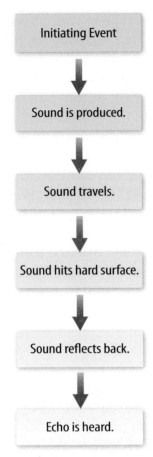

Figure 4 Events-chain concept maps show the order of steps in a process or event. This concept map shows how a sound makes an echo.

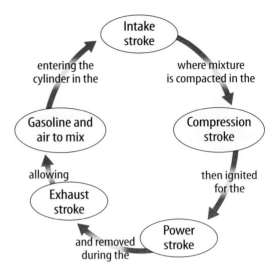

Figure 5 A cycle map shows events that occur in a cycle.

Cycle Map A specific type of events chain is a cycle map. It is used when the series of events do not produce a final outcome, but instead relate back to the beginning event, such as in **Figure 5.** Therefore, the cycle repeats itself.

To make a cycle map, first decide what event is the beginning event. This is also called the initiating event. Then list the next events in the order that they occur, with the last event relating back to the initiating event. Words can be written between the events that describe what happens from one event to the next. The number of events in a cycle map can vary, but usually contain three or more events.

Spider Map A type of concept map that you can use for brainstorming is the spider map. When you have a central idea, you might find that you have a jumble of ideas that relate to it but are not necessarily clearly related to each other. The spider map on sound in **Figure 6** shows that if you write these ideas outside the main concept, then you can begin to separate and group unrelated terms so they become more useful.

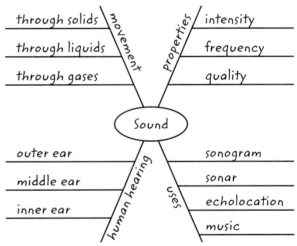

Figure 6 A spider map allows you to list ideas that relate to a central topic but not necessarily to one another.

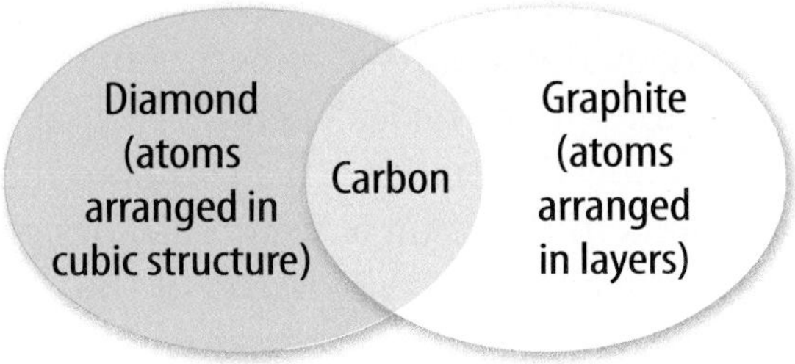

Figure 7 This Venn diagram compares and contrasts two substances made from carbon.

Venn Diagram To illustrate how two subjects compare and contrast you can use a Venn diagram. You can see the characteristics that the subjects have in common and those that they do not, shown in **Figure 7.**

To create a Venn diagram, draw two overlapping ovals that are big enough to write in. List the characteristics unique to one subject in one oval, and the characteristics of the other subject in the other oval. The characteristics in common are listed in the overlapping section.

Make and Use Tables One way to organize information so it is easier to understand is to use a table. Tables can contain numbers, words, or both.

To make a table, list the items to be compared in the first column and the characteristics to be compared in the first row. The title should clearly indicate the content of the table, and the column or row heads should be clear. Notice that in **Table 1** the units are included.

Table 1 Recyclables Collected During Week			
Day of Week	Paper (kg)	Aluminum (kg)	Glass (kg)
Monday	5.0	4.0	12.0
Wednesday	4.0	1.0	10.0
Friday	2.5	2.0	10.0

Make a Model One way to help you better understand the parts of a structure, the way a process works, or to show things too large or small for viewing is to make a model. For example, an atomic model made of a plastic-ball nucleus and chenille stem electron shells can help you visualize how the parts of an atom relate to each other. Other types of models can be devised on a computer or represented by equations.

Form a Hypothesis

A possible explanation based on previous knowledge and observations is called a hypothesis. After researching gasoline types and recalling previous experiences in your family's car you form a hypothesis—our car runs more efficiently because we use premium gasoline. To be valid, a hypothesis has to be something you can test by using an investigation.

Predict When you apply a hypothesis to a specific situation, you predict something about that situation. A prediction makes a statement in advance, based on prior observation, experience, or scientific reasoning. People use predictions to make everyday decisions. Scientists test predictions by performing investigations. Based on previous observations and experiences, you might form a prediction that cars are more efficient with premium gasoline. The prediction can be tested in an investigation.

Design an Experiment A scientist needs to make many decisions before beginning an investigation. Some of these include: how to carry out the investigation, what steps to follow, how to record the data, and how the investigation will answer the question. It also is important to address any safety concerns.

SCIENCE SKILL HANDBOOK

MATH SKILL HANDBOOK

FOLDABLES HANDBOOK

REFERENCE HANDBOOK

GLOSSARY/ GLOSARIO

INDEX

Test the Hypothesis

Now that you have formed your hypothesis, you need to test it. Using an investigation, you will make observations and collect data, or information. This data might either support or not support your hypothesis. Scientists collect and organize data as numbers and descriptions.

Follow a Procedure In order to know what materials to use, as well as how and in what order to use them, you must follow a procedure. **Figure 8** shows a procedure you might follow to test your hypothesis.

Procedure	
Step 1	Use regular gasoline for two weeks.
Step 2	Record the number of kilometers between fill-ups and the amount of gasoline used.
Step 3	Switch to premium gasoline for two weeks.
Step 4	Record the number of kilometers between fill-ups and the amount of gasoline used.

Figure 8 A procedure tells you what to do step-by-step.

Identify and Manipulate Variables and Controls In any experiment, it is important to keep everything the same except for the item you are testing. The one factor you change is called the independent variable. The change that results is the dependent variable. Make sure you have only one independent variable, to assure yourself of the cause of the changes you observe in the dependent variable. For example, in your gasoline experiment the type of fuel is the independent variable. The dependent variable is the efficiency.

Many experiments also have a control—an individual instance or experimental subject for which the independent variable is not changed. You can then compare the test results to the control results. To design a control you can have two cars of the same type. The control car uses regular gasoline for four weeks. After you are done with the test, you can compare the experimental results to the control results.

Collect Data

Whether you are carrying out an investigation or a short observational experiment, you will collect data, as shown in **Figure 9.** Scientists collect data as numbers and descriptions and organize them in specific ways.

Observe Scientists observe items and events, then record what they see. When they use only words to describe an observation, it is called qualitative data. Scientists' observations also can describe how much there is of something. These observations use numbers, as well as words, in the description and are called quantitative data. For example, if a sample of the element gold is described as being "shiny and very dense" the data are qualitative. Quantitative data on this sample of gold might include "a mass of 30 g and a density of 19.3 g/cm^3."

Figure 9 Collecting data is one way to gather information directly.

Figure 10 Record data neatly and clearly so it is easy to understand.

When you make observations you should examine the entire object or situation first, and then look carefully for details. It is important to record observations accurately and completely. Always record your notes immediately as you make them, so you do not miss details or make a mistake when recording results from memory. Never put unidentified observations on scraps of paper. Instead they should be recorded in a notebook, like the one in **Figure 10.** Write your data neatly so you can easily read it later. At each point in the experiment, record your observations and label them. That way, you will not have to determine what the figures mean when you look at your notes later. Set up any tables that you will need to use ahead of time, so you can record any observations right away. Remember to avoid bias when collecting data by not including personal thoughts when you record observations. Record only what you observe.

Estimate Scientific work also involves estimating. To estimate is to make a judgment about the size or the number of something without measuring or counting. This is important when the number or size of an object or population is too large or too difficult to accurately count or measure.

Sample Scientists may use a sample or a portion of the total number as a type of estimation. To sample is to take a small, representative portion of the objects or organisms of a population for research. By making careful observations or manipulating variables within that portion of the group, information is discovered and conclusions are drawn that might apply to the whole population. A poorly chosen sample can be unrepresentative of the whole. If you were trying to determine the rainfall in an area, it would not be best to take a rainfall sample from under a tree.

Measure You use measurements every day. Scientists also take measurements when collecting data. When taking measurements, it is important to know how to use measuring tools properly. Accuracy also is important.

Length To measure length, the distance between two points, scientists use meters. Smaller measurements might be measured in centimeters or millimeters.

Length is measured using a metric ruler or meterstick. When using a metric ruler, line up the 0-cm mark with the end of the object being measured and read the number of the unit where the object ends. Look at the metric ruler shown in **Figure 11.** The centimeter lines are the long, numbered lines, and the shorter lines are millimeter lines. In this instance, the length would be 4.50 cm.

Figure 11 This metric ruler has centimeter and millimeter divisions.

SCIENCE SKILL HANDBOOK

MATH SKILL HANDBOOK

FOLDABLES HANDBOOK

REFERENCE HANDBOOK

GLOSSARY/ GLOSARIO

INDEX

Science Skill Handbook

Math Skill Handbook

Foldables Handbook

Reference Handbook

Glossary/ Glosario

Index

Mass The SI unit for mass is the kilogram (kg). Scientists can measure mass using units formed by adding metric prefixes to the unit gram (g), such as milligram (mg). To measure mass, you might use a triple-beam balance similar to the one shown in **Figure 12.** The balance has a pan on one side and a set of beams on the other side. Each beam has a rider that slides on the beam.

When using a triple-beam balance, place an object on the pan. Slide the largest rider along its beam until the pointer drops below zero. Then move it back one notch. Repeat the process for each rider proceeding from the larger to smaller until the pointer swings an equal distance above and below the zero point. Sum the masses on each beam to find the mass of the object. Move all riders back to zero when finished.

Instead of putting materials directly on the balance, scientists often take a tare of a container. A tare is the mass of a container into which objects or substances are placed for measuring their masses. To find the mass of objects or substances, find the mass of a clean container. Remove the container from the pan, and place the object or substances in the container. Find the mass of the container with the materials in it. Subtract the mass of the empty container from the mass of the filled container to find the mass of the materials you are using.

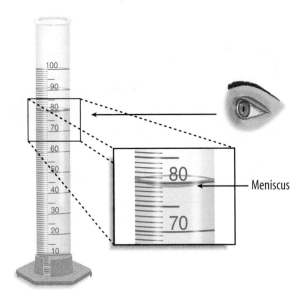

Figure 13 Graduated cylinders measure liquid volume.

Liquid Volume To measure liquids, the unit used is the liter. When a smaller unit is needed, scientists might use a milliliter. Because a milliliter takes up the volume of a cube measuring 1 cm on each side it also can be called a cubic centimeter ($cm^3 = cm \times cm \times cm$).

You can use beakers and graduated cylinders to measure liquid volume. A graduated cylinder, shown in **Figure 13,** is marked from bottom to top in milliliters. In lab, you might use a 10-mL graduated cylinder or a 100-mL graduated cylinder. When measuring liquids, notice that the liquid has a curved surface. Look at the surface at eye level, and measure the bottom of the curve. This is called the meniscus. The graduated cylinder in **Figure 13** contains 79.0 mL, or 79.0 cm^3, of a liquid.

Temperature Scientists often measure temperature using the Celsius scale. Pure water has a freezing point of 0°C and boiling point of 100°C. The unit of measurement is degrees Celsius. Two other scales often used are the Fahrenheit and Kelvin scales.

Figure 12 A triple-beam balance is used to determine the mass of an object.

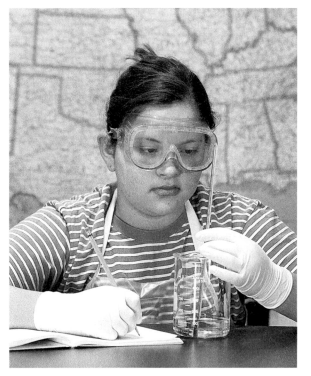

Figure 14 A thermometer measures the temperature of an object.

Scientists use a thermometer to measure temperature. Most thermometers in a laboratory are glass tubes with a bulb at the bottom end containing a liquid such as colored alcohol. The liquid rises or falls with a change in temperature. To read a glass thermometer like the thermometer in **Figure 14,** rotate it slowly until a red line appears. Read the temperature where the red line ends.

Form Operational Definitions An operational definition defines an object by how it functions, works, or behaves. For example, when you are playing hide and seek and a tree is home base, you have created an operational definition for a tree.

Objects can have more than one operational definition. For example, a ruler can be defined as a tool that measures the length of an object (how it is used). It can also be a tool with a series of marks used as a standard when measuring (how it works).

Analyze the Data

To determine the meaning of your observations and investigation results, you will need to look for patterns in the data. Then you must think critically to determine what the data mean. Scientists use several approaches when they analyze the data they have collected and recorded. Each approach is useful for identifying specific patterns.

Interpret Data The word *interpret* means "to explain the meaning of something." When analyzing data from an experiment, try to find out what the data show. Identify the control group and the test group to see whether changes in the independent variable have had an effect. Look for differences in the dependent variable between the control and test groups.

Classify Sorting objects or events into groups based on common features is called classifying. When classifying, first observe the objects or events to be classified. Then select one feature that is shared by some members in the group, but not by all. Place those members that share that feature in a subgroup. You can classify members into smaller and smaller subgroups based on characteristics. Remember that when you classify, you are grouping objects or events for a purpose. Keep your purpose in mind as you select the features to form groups and subgroups.

Compare and Contrast Observations can be analyzed by noting the similarities and differences between two or more objects or events that you observe. When you look at objects or events to see how they are similar, you are comparing them. Contrasting is looking for differences in objects or events.

SCIENCE SKILL HANDBOOK

MATH SKILL HANDBOOK

FOLDABLES HANDBOOK

REFERENCE HANDBOOK

GLOSSARY/ GLOSARIO

INDEX

SCIENCE SKILL HANDBOOK

MATH SKILL HANDBOOK

FOLDABLES HANDBOOK

REFERENCE HANDBOOK

GLOSSARY/ GLOSARIO

INDEX

Recognize Cause and Effect A cause is a reason for an action or condition. The effect is that action or condition. When two events happen together, it is not necessarily true that one event caused the other. Scientists must design a controlled investigation to recognize the exact cause and effect.

Draw Conclusions

When scientists have analyzed the data they collected, they proceed to draw conclusions about the data. These conclusions are sometimes stated in words similar to the hypothesis that you formed earlier. They may confirm a hypothesis, or lead you to a new hypothesis.

Infer Scientists often make inferences based on their observations. An inference is an attempt to explain observations or to indicate a cause. An inference is not a fact, but a logical conclusion that needs further investigation. For example, you may infer that a fire has caused smoke. Until you investigate, however, you do not know for sure.

Apply When you draw a conclusion, you must apply those conclusions to determine whether the data supports the hypothesis. If your data do not support your hypothesis, it does not mean that the hypothesis is wrong. It means only that the result of the investigation did not support the hypothesis. Maybe the experiment needs to be redesigned, or some of the initial observations on which the hypothesis was based were incomplete or biased. Perhaps more observation or research is needed to refine your hypothesis. A successful investigation does not always come out the way you originally predicted.

Avoid Bias Sometimes a scientific investigation involves making judgments. When you make a judgment, you form an opinion. It is important to be honest and not to allow any expectations of results to bias your judgments. This is important throughout the entire investigation, from researching to collecting data to drawing conclusions.

Communicate

The communication of ideas is an important part of the work of scientists. A discovery that is not reported will not advance the scientific community's understanding or knowledge. Communication among scientists also is important as a way of improving their investigations.

Scientists communicate in many ways, from writing articles in journals and magazines that explain their investigations and experiments, to announcing important discoveries on television and radio. Scientists also share ideas with colleagues on the Internet or present them as lectures, like the student is doing in **Figure 15.**

Figure 15 A student communicates to his peers about his investigation.

These safety symbols are used in laboratory and field investigations in this book to indicate possible hazards. Learn the meaning of each symbol and refer to this page often. *Remember to wash your hands thoroughly after completing lab procedures.*

PROTECTIVE EQUIPMENT   Do not begin any lab without the proper protection equipment.

GOGGLES	Proper eye protection must be worn when performing or observing science activities that involve items or conditions as listed below.	**APRON** Wear an approved apron when using substances that could stain, wet, or destroy cloth.	**SOAP** Wash hands with soap and water before removing goggles and after all lab activities.

 GLOVES Wear gloves when working with biological materials, chemicals, animals, or materials that can stain or irritate hands.

LABORATORY HAZARDS

Symbols	Potential Hazards	Precaution	Response
DISPOSAL	contamination of classroom or environment due to improper disposal of materials such as chemicals and live specimens	• DO NOT dispose of hazardous materials in the sink or trash can. • Dispose of wastes as directed by your teacher.	• If hazardous materials are disposed of improperly, notify your teacher immediately.
EXTREME TEMPERATURE	skin burns due to extremely hot or cold materials such as hot glass, liquids, or metals; liquid nitrogen; dry ice	• Use proper protective equipment, such as hot mitts and/or tongs, when handling objects with extreme temperatures.	• If injury occurs, notify your teacher immediately.
SHARP OBJECTS	punctures or cuts from sharp objects such as razor blades, pins, scalpels, and broken glass	• Handle glassware carefully to avoid breakage. • Walk with sharp objects pointed downward, away from you and others.	• If broken glass or injury occurs, notify your teacher immediately.
ELECTRICAL	electric shock or skin burn due to improper grounding, short circuits, liquid spills, or exposed wires	• Check condition of wires and apparatus for fraying or uninsulated wires, and broken or cracked equipment. • Use only GFCI-protected outlets	• DO NOT attempt to fix electrical problems. Notify your teacher immediately.
CHEMICAL	skin irritation or burns, breathing difficulty, and/or poisoning due to touching, swallowing, or inhalation of chemicals such as acids, bases, bleach, metal compounds, iodine, poinsettias, pollen, ammonia, acetone, nail polish remover, heated chemicals, mothballs, and any other chemicals labeled or known to be dangerous	• Wear proper protective equipment such as goggles, apron, and gloves when using chemicals. • Ensure proper room ventilation or use a fume hood when using materials that produce fumes. • NEVER smell fumes directly. • NEVER taste or eat any material in the laboratory.	• If contact occurs, immediately flush affected area with water and notify your teacher. • If a spill occurs, leave the area immediately and notify your teacher.
FLAMMABLE	unexpected fire due to liquids or gases that ignite easily such as rubbing alcohol	• Avoid open flames, sparks, or heat when flammable liquids are present.	• If a fire occurs, leave the area immediately and notify your teacher.
OPEN FLAME	burns or fire due to open flame from matches, Bunsen burners, or burning materials	• Tie back loose hair and clothing. • Keep flame away from all materials. • Follow teacher instructions when lighting and extinguishing flames. • Use proper protection, such as hot mitts or tongs, when handling hot objects.	• If a fire occurs, leave the area immediately and notify your teacher.
ANIMAL SAFETY	injury to or from laboratory animals	• Wear proper protective equipment such as gloves, apron, and goggles when working with animals. • Wash hands after handling animals.	• If injury occurs, notify your teacher immediately.
BIOLOGICAL	infection or adverse reaction due to contact with organisms such as bacteria, fungi, and biological materials such as blood, animal or plant materials	• Wear proper protective equipment such as gloves, goggles, and apron when working with biological materials. • Avoid skin contact with an organism or any part of the organism. • Wash hands after handling organisms.	• If contact occurs, wash the affected area and notify your teacher immediately.
FUME	breathing difficulties from inhalation of fumes from substances such as ammonia, acetone, nail polish remover, heated chemicals, and mothballs	• Wear goggles, apron, and gloves. • Ensure proper room ventilation or use a fume hood when using substances that produce fumes. • NEVER smell fumes directly.	• If a spill occurs, leave area and notify your teacher immediately.
IRRITANT	irritation of skin, mucous membranes, or respiratory tract due to materials such as acids, bases, bleach, pollen, mothballs, steel wool, and potassium permanganate	• Wear goggles, apron, and gloves. • Wear a dust mask to protect against fine particles.	• If skin contact occurs, immediately flush the affected area with water and notify your teacher.
RADIOACTIVE	excessive exposure from alpha, beta, and gamma particles	• Remove gloves and wash hands with soap and water before removing remainder of protective equipment.	• If cracks or holes are found in the container, notify your teacher immediately.

SCIENCE SKILL HANDBOOK

MATH SKILL HANDBOOK

FOLDABLES HANDBOOK

REFERENCE HANDBOOK

GLOSSARY/ GLOSARIO

INDEX

Safety in the Science Laboratory

Introduction to Science Safety

The science laboratory is a safe place to work if you follow standard safety procedures. Being responsible for your own safety helps to make the entire laboratory a safer place for everyone. When performing any lab, read and apply the caution statements and safety symbol listed at the beginning of the lab.

General Safety Rules

1. Complete the *Lab Safety Form* or other safety contract BEFORE starting any science lab.

2. Study the procedure. Ask your teacher any questions. Be sure you understand safety symbols shown on the page.

3. Notify your teacher about allergies or other health conditions that can affect your participation in a lab.

4. Learn and follow use and safety procedures for your equipment. If unsure, ask your teacher.

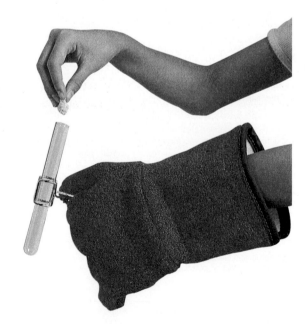

5. Never eat, drink, chew gum, apply cosmetics, or do any personal grooming in the lab. Never use lab glassware as food or drink containers. Keep your hands away from your face and mouth.

6. Know the location and proper use of the safety shower, eye wash, fire blanket, and fire alarm.

Prevent Accidents

1. Use the safety equipment provided to you. Goggles and a safety apron should be worn during investigations.

2. Do NOT use hair spray, mousse, or other flammable hair products. Tie back long hair and tie down loose clothing.

3. Do NOT wear sandals or other open-toed shoes in the lab.

4. Remove jewelry on hands and wrists. Loose jewelry, such as chains and long necklaces, should be removed to prevent them from getting caught in equipment.

5. Do not taste any substances or draw any material into a tube with your mouth.

6. Proper behavior is expected in the lab. Practical jokes and fooling around can lead to accidents and injury.

7. Keep your work area uncluttered.

Laboratory Work

1. Collect and carry all equipment and materials to your work area before beginning a lab.

2. Remain in your own work area unless given permission by your teacher to leave it.

SCIENCE SKILL HANDBOOK

MATH SKILL HANDBOOK

FOLDABLES HANDBOOK

REFERENCE HANDBOOK

GLOSSARY/ GLOSARIO

INDEX

3. Always slant test tubes away from your-self and others when heating them, adding substances to them, or rinsing them.

4. If instructed to smell a substance in a container, hold the container a short distance away and fan vapors toward your nose.

5. Do NOT substitute other chemicals/substances for those in the materials list unless instructed to do so by your teacher.

6. Do NOT take any materials or chemicals outside of the laboratory.

7. Stay out of storage areas unless instructed to be there and supervised by your teacher.

Laboratory Cleanup

1. Turn off all burners, water, and gas, and disconnect all electrical devices.

2. Clean all pieces of equipment and return all materials to their proper places.

3. Dispose of chemicals and other materials as directed by your teacher. Place broken glass and solid substances in the proper containers. Never discard materials in the sink.

4. Clean your work area.

5. Wash your hands with soap and water thoroughly BEFORE removing your goggles.

Emergencies

1. Report any fire, electrical shock, glass-ware breakage, spill, or injury, no matter how small, to your teacher immediately. Follow his or her instructions.

2. If your clothing should catch fire, STOP, DROP, and ROLL. If possible, smother it with the fire blanket or get under a safety shower. NEVER RUN.

3. If a fire should occur, turn off all gas and leave the room according to estab-lished procedures.

4. In most instances, your teacher will clean up spills. Do NOT attempt to clean up spills unless you are given permission and instructions to do so.

5. If chemicals come into contact with your eyes or skin, notify your teacher immediately. Use the eyewash, or flush your skin or eyes with large quantities of water.

6. The fire extinguisher and first-aid kit should only be used by your teacher unless it is an extreme emergency and you have been given permission.

7. If someone is injured or becomes ill, only a professional medical provider or someone certified in first aid should perform first-aid procedures.

SCIENCE SKILL HANDBOOK

MATH SKILL HANDBOOK

FOLDABLES HANDBOOK

REFERENCE HANDBOOK

GLOSSARY/ GLOSARIO

INDEX

SCIENCE SKILL HANDBOOK

MATH SKILL HANDBOOK

FOLDABLES HANDBOOK

REFERENCE HANDBOOK

GLOSSARY/ GLOSARIO

INDEX

Use Fractions

A fraction compares a part to a whole. In the fraction $\frac{2}{3}$, the 2 represents the part and is the numerator. The 3 represents the whole and is the denominator.

Reduce Fractions To reduce a fraction, you must find the largest factor that is common to both the numerator and the denominator, the greatest common factor (GCF). Divide both numbers by the GCF. The fraction has then been reduced, or it is in its simplest form.

Example

Twelve of the 20 chemicals in the science lab are in powder form. What fraction of the chemicals used in the lab are in powder form?

Step 1 Write the fraction.

$$\frac{part}{whole} = \frac{12}{20}$$

Step 2 To find the GCF of the numerator and denominator, list all of the factors of each number.

Factors of 12: 1, 2, 3, 4, 6, 12 (the numbers that divide evenly into 12)

Factors of 20: 1, 2, 4, 5, 10, 20 (the numbers that divide evenly into 20)

Step 3 List the common factors.

1, 2, 4

Step 4 Choose the greatest factor in the list. The GCF of 12 and 20 is 4.

Step 5 Divide the numerator and denominator by the GCF.

$$\frac{12 \div 4}{20 \div 4} = \frac{3}{5}$$

In the lab, $\frac{3}{5}$ of the chemicals are in powder form.

Practice Problem At an amusement park, 66 of 90 rides have a height restriction. What fraction of the rides, in its simplest form, has a height restriction?

Add and Subtract Fractions with Like Denominators To add or subtract fractions with the same denominator, add or subtract the numerators and write the sum or difference over the denominator. After finding the sum or difference, find the simplest form for your fraction.

Example 1

In the forest outside your house, $\frac{1}{8}$ of the animals are rabbits, $\frac{3}{8}$ are squirrels, and the remainder are birds and insects. How many are mammals?

Step 1 Add the numerators.

$$\frac{1}{8} + \frac{3}{8} = \frac{(1 + 3)}{8} = \frac{4}{8}$$

Step 2 Find the GCF.

$$\frac{4}{8} \text{ (GCF, 4)}$$

Step 3 Divide the numerator and denominator by the GCF.

$$\frac{4 \div 4}{8 \div 4} = \frac{1}{2}$$

$\frac{1}{2}$ of the animals are mammals.

Example 2

If $\frac{7}{16}$ of the Earth is covered by freshwater, and $\frac{1}{16}$ of that is in glaciers, how much freshwater is not frozen?

Step 1 Subtract the numerators.

$$\frac{7}{16} - \frac{1}{16} = \frac{(7 - 1)}{16} = \frac{6}{16}$$

Step 2 Find the GCF.

$$\frac{6}{16} \text{ (GCF, 2)}$$

Step 3 Divide the numerator and denominator by the GCF.

$$\frac{6 \div 2}{16 \div 2} = \frac{3}{8}$$

$\frac{3}{8}$ of the freshwater is not frozen.

Practice Problem A bicycle rider is riding at a rate of 15 km/h for $\frac{4}{9}$ of his ride, 10 km/h for $\frac{2}{9}$ of his ride, and 8 km/h for the remainder of the ride. How much of his ride is he riding at a rate greater than 8 km/h?

Add and Subtract Fractions with Unlike Denominators To add or subtract fractions with unlike denominators, first find the least common denominator (LCD). This is the smallest number that is a common multiple of both denominators. Rename each fraction with the LCD, and then add or subtract. Find the simplest form if necessary.

Example 1

A chemist makes a paste that is $\frac{1}{2}$ table salt (NaCl), $\frac{1}{3}$ sugar ($C_6H_{12}O_6$), and the remainder is water (H_2O). How much of the paste is a solid?

Step 1 Find the LCD of the fractions.

$$\frac{1}{2} + \frac{1}{3} \text{ (LCD, 6)}$$

Step 2 Rename each numerator and each denominator with the LCD.

Step 3 Add the numerators.

$$\frac{3}{6} + \frac{2}{6} = \frac{(3+2)}{6} = \frac{5}{6}$$

$\frac{5}{6}$ of the paste is a solid.

Example 2

The average precipitation in Grand Junction, CO, is $\frac{7}{10}$ inch in November, and $\frac{3}{5}$ inch in December. What is the total average precipitation?

Step 1 Find the LCD of the fractions.

$$\frac{7}{10} + \frac{3}{5} \text{ (LCD, 10)}$$

Step 2 Rename each numerator and each denominator with the LCD.

Step 3 Add the numerators.

$$\frac{7}{10} + \frac{6}{10} = \frac{(7+6)}{10} = \frac{13}{10}$$

$\frac{13}{10}$ inches total precipitation, or $1\frac{3}{10}$ inches.

Practice Problem On an electric bill, about $\frac{1}{8}$ of the energy is from solar energy and about $\frac{1}{10}$ is from wind power. How much of the total bill is from solar energy and wind power combined?

Example 3

In your body, $\frac{7}{10}$ of your muscle contractions are involuntary (cardiac and smooth muscle tissue). Smooth muscle makes $\frac{3}{15}$ of your muscle contractions. How many of your muscle contractions are made by cardiac muscle?

Step 1 Find the LCD of the fractions.

$$\frac{7}{10} - \frac{3}{15} \text{ (LCD, 30)}$$

Step 2 Rename each numerator and each denominator with the LCD.

$$\frac{7 \times 3}{10 \times 3} = \frac{21}{30}$$

$$\frac{3 \times 2}{15 \times 2} = \frac{6}{30}$$

Step 3 Subtract the numerators.

$$\frac{21}{30} - \frac{6}{30} = \frac{(21-6)}{30} = \frac{15}{30}$$

Step 4 Find the GCF.

$$\frac{15}{30} \text{ (GCF, 15)}$$

$$\frac{1}{2}$$

$\frac{1}{2}$ of all muscle contractions are cardiac muscle.

Example 4

Tony wants to make cookies that call for $\frac{3}{4}$ of a cup of flour, but he only has $\frac{1}{3}$ of a cup. How much more flour does he need?

Step 1 Find the LCD of the fractions.

$$\frac{3}{4} - \frac{1}{3} \text{ (LCD, 12)}$$

Step 2 Rename each numerator and each denominator with the LCD.

$$\frac{3 \times 3}{4 \times 3} = \frac{9}{12}$$

$$\frac{1 \times 4}{3 \times 4} = \frac{4}{12}$$

Step 3 Subtract the numerators.

$$\frac{9}{12} - \frac{4}{12} = \frac{(9-4)}{12} = \frac{5}{12}$$

$\frac{5}{12}$ of a cup of flour

Practice Problem Using the information provided to you in Example 3 above, determine how many muscle contractions are voluntary (skeletal muscle).

Multiply Fractions To multiply with fractions, multiply the numerators and multiply the denominators. Find the simplest form if necessary.

Example

Multiply $\frac{3}{5}$ by $\frac{1}{3}$.

Step 1 Multiply the numerators and denominators.

$$\frac{3}{5} \times \frac{1}{3} = \frac{(3 \times 1)}{(5 \times 3)} \; \frac{3}{15}$$

Step 2 Find the GCF.

$$\frac{3}{15} \; (\text{GCF, 3})$$

Step 3 Divide the numerator and denominator by the GCF.

$$\frac{3 \div 3}{15 \div 3} = \frac{1}{5}$$

$\frac{3}{5}$ multiplied by $\frac{1}{3}$ is $\frac{1}{5}$.

Practice Problem Multiply $\frac{3}{14}$ by $\frac{5}{16}$.

Find a Reciprocal Two numbers whose product is 1 are called multiplicative inverses, or reciprocals.

Example

Find the reciprocal of $\frac{3}{8}$.

Step 1 Inverse the fraction by putting the denominator on top and the numerator on the bottom.

$$\frac{8}{3}$$

The reciprocal of $\frac{3}{8}$ is $\frac{8}{3}$.

Practice Problem Find the reciprocal of $\frac{4}{9}$.

Divide Fractions To divide one fraction by another fraction, multiply the dividend by the reciprocal of the divisor. Find the simplest form if necessary.

Example 1

Divide $\frac{1}{9}$ by $\frac{1}{3}$.

Step 1 Find the reciprocal of the divisor.

The reciprocal of $\frac{1}{3}$ is $\frac{3}{1}$.

Step 2 Multiply the dividend by the reciprocal of the divisor.

$$\frac{\frac{1}{9}}{\frac{1}{3}} = \frac{1}{9} \times \frac{3}{1} = \frac{(1 \times 3)}{(9 \times 1)} = \frac{3}{9}$$

Step 3 Find the GCF.

$$\frac{3}{9} \; (\text{GCF, 3})$$

Step 4 Divide the numerator and denominator by the GCF.

$$\frac{3 \div 3}{9 \div 3} = \frac{1}{3}$$

$\frac{1}{9}$ divided by $\frac{1}{3}$ is $\frac{1}{3}$.

Example 2

Divide $\frac{3}{5}$ by $\frac{1}{4}$.

Step 1 Find the reciprocal of the divisor.

The reciprocal of $\frac{1}{4}$ is $\frac{4}{1}$.

Step 2 Multiply the dividend by the reciprocal of the divisor.

$$\frac{\frac{3}{5}}{\frac{1}{4}} = \frac{3}{5} \times \frac{4}{1} = \frac{(3 \times 4)}{(5 \times 1)} = \frac{12}{5}$$

$\frac{3}{5}$ divided by $\frac{1}{4}$ is $\frac{12}{5}$ or $2\frac{2}{5}$.

Practice Problem Divide $\frac{3}{11}$ by $\frac{7}{10}$.

SCIENCE SKILL HANDBOOK

MATH SKILL HANDBOOK

FOLDABLES HANDBOOK

REFERENCE HANDBOOK

GLOSSARY/ GLOSARIO

INDEX

Use Ratios

When you compare two numbers by division, you are using a ratio. Ratios can be written 3 to 5, 3:5, or $\frac{3}{5}$. Ratios, like fractions, also can be written in simplest form.

Ratios can represent one type of probability, called odds. This is a ratio that compares the number of ways a certain outcome occurs to the number of possible outcomes. For example, if you flip a coin 100 times, what are the odds that it will come up heads? There are two possible outcomes, heads or tails, so the odds of coming up heads are 50:100. Another way to say this is that 50 out of 100 times the coin will come up heads. In its simplest form, the ratio is 1:2.

Example 1

A chemical solution contains 40 g of salt and 64 g of baking soda. What is the ratio of salt to baking soda as a fraction in simplest form?

Step 1 Write the ratio as a fraction.

$$\frac{\text{salt}}{\text{baking soda}} = \frac{40}{64}$$

Step 2 Express the fraction in simplest form. The GCF of 40 and 64 is 8.

$$\frac{40}{64} = \frac{40 \div 8}{64 \div 8} = \frac{5}{8}$$

The ratio of salt to baking soda in the sample is 5:8.

Example 2

Sean rolls a 6-sided die 6 times. What are the odds that the side with a 3 will show?

Step 1 Write the ratio as a fraction.

$$\frac{\text{number of sides with a 3}}{\text{number of possible sides}} = \frac{1}{6}$$

Step 2 Multiply by the number of attempts.

$$\frac{1}{6} \times 6 \text{ attempts} = \frac{6}{6} \text{ attempts} = 1 \text{ attempt}$$

1 attempt out of 6 will show a 3.

Practice Problem Two metal rods measure 100 cm and 144 cm in length. What is the ratio of their lengths in simplest form?

Use Decimals

A fraction with a denominator that is a power of ten can be written as a decimal. For example, 0.27 means $\frac{27}{100}$. The decimal point separates the ones place from the tenths place.

Any fraction can be written as a decimal using division. For example, the fraction $\frac{5}{8}$ can be written as a decimal by dividing 5 by 8. Written as a decimal, it is 0.625.

Add or Subtract Decimals When adding and subtracting decimals, line up the decimal points before carrying out the operation.

Example 1

Find the sum of 47.68 and 7.80.

Step 1 Line up the decimal places when you write the numbers.

$$\begin{array}{r} 47.68 \\ + 7.80 \end{array}$$

Step 2 Add the decimals.

$$\begin{array}{r} \overset{1\ 1}{47.68} \\ + 7.80 \\ \hline 55.48 \end{array}$$

The sum of 47.68 and 7.80 is 55.48.

Example 2

Find the difference of 42.17 and 15.85.

Step 1 Line up the decimal places when you write the number.

$$\begin{array}{r} 42.17 \\ -15.85 \end{array}$$

Step 2 Subtract the decimals.

$$\begin{array}{r} \overset{3\ 11}{4\cancel{2}.17} \\ -15.85 \\ \hline 26.32 \end{array}$$

The difference of 42.17 and 15.85 is 26.32.

Practice Problem Find the sum of 1.245 and 3.842.

SCIENCE SKILL HANDBOOK

MATH SKILL HANDBOOK

FOLDABLES HANDBOOK

REFERENCE HANDBOOK

GLOSSARY/ GLOSARIO

INDEX

SCIENCE SKILL HANDBOOK

MATH SKILL HANDBOOK

FOLDABLES HANDBOOK

REFERENCE HANDBOOK

GLOSSARY/ GLOSARIO

INDEX

Multiply Decimals To multiply decimals, multiply the numbers like numbers without decimal points. Count the decimal places in each factor. The product will have the same number of decimal places as the sum of the decimal places in the factors.

Example

Multiply 2.4 by 5.9.

Step 1 Multiply the factors like two whole numbers.

$24 \times 59 = 1416$

Step 2 Find the sum of the number of decimal places in the factors. Each factor has one decimal place, for a sum of two decimal places.

Step 3 The product will have two decimal places.

14.16

The product of 2.4 and 5.9 is 14.16.

Practice Problem Multiply 4.6 by 2.2.

Divide Decimals When dividing decimals, change the divisor to a whole number. To do this, multiply both the divisor and the dividend by the same power of ten. Then place the decimal point in the quotient directly above the decimal point in the dividend. Then divide as you do with whole numbers.

Example

Divide 8.84 by 3.4.

Step 1 Multiply both factors by 10.

$3.4 \times 10 = 34, 8.84 \times 10 = 88.4$

Step 2 Divide 88.4 by 34.

```
        2.6
   34)88.4
      −68
      204
     −204
        0
```

8.84 divided by 3.4 is 2.6.

Practice Problem Divide 75.6 by 3.6.

Use Proportions

An equation that shows that two ratios are equivalent is a proportion. The ratios $\frac{2}{4}$ and $\frac{5}{10}$ are equivalent, so they can be written as $\frac{2}{4} = \frac{5}{10}$. This equation is a proportion.

When two ratios form a proportion, the cross products are equal. To find the cross products in the proportion $\frac{2}{4} = \frac{5}{10}$, multiply the 2 and the 10, and the 4 and the 5. Therefore $2 \times 10 = 4 \times 5$, or $20 = 20$.

Because you know that both ratios are equal, you can use cross products to find a missing term in a proportion. This is known as solving the proportion.

Example

The heights of a tree and a pole are proportional to the lengths of their shadows. The tree casts a shadow of 24 m when a 6-m pole casts a shadow of 4 m. What is the height of the tree?

Step 1 Write a proportion.

$$\frac{\text{height of tree}}{\text{height of pole}} = \frac{\text{length of tree's shadow}}{\text{length of pole's shadow}}$$

Step 2 Substitute the known values into the proportion. Let h represent the unknown value, the height of the tree.

$$\frac{h}{6} \times \frac{24}{4}$$

Step 3 Find the cross products.

$h \times 4 = 6 \times 24$

Step 4 Simplify the equation.

$4h \times 144$

Step 5 Divide each side by 4.

$$\frac{4h}{4} \times \frac{144}{4}$$

$h = 36$

The height of the tree is 36 m.

Practice Problem The ratios of the weights of two objects on the Moon and on Earth are in proportion. A rock weighing 3 N on the Moon weighs 18 N on Earth. How much would a rock that weighs 5 N on the Moon weigh on Earth?

Use Percentages

The word *percent* means "out of one hundred." It is a ratio that compares a number to 100. Suppose you read that 77 percent of Earth's surface is covered by water. That is the same as reading that the fraction of Earth's surface covered by water is $\frac{77}{100}$. To express a fraction as a percent, first find the equivalent decimal for the fraction. Then, multiply the decimal by 100 and add the percent symbol.

Example 1

Express $\frac{13}{20}$ as a percent.

Step 1 Find the equivalent decimal for the fraction.

$$
\begin{array}{r}
0.65 \\
20\overline{)13.00} \\
\underline{12\,0} \\
1\,00 \\
\underline{1\,00} \\
0
\end{array}
$$

Step 2 Rewrite the fraction $\frac{13}{20}$ as 0.65.

Step 3 Multiply 0.65 by 100 and add the % symbol.

$$0.65 \times 100 = 65 = 65\%$$

So, $\frac{13}{20} = 65\%$.

This also can be solved as a proportion.

Example 2

Express $\frac{13}{20}$ as a percent.

Step 1 Write a proportion.

$$\frac{13}{20} = \frac{x}{100}$$

Step 2 Find the cross products.

$$1300 = 20x$$

Step 3 Divide each side by 20.

$$\frac{1300}{20} = \frac{20x}{20}$$

$$65\% = x$$

Practice Problem In one year, 73 of 365 days were rainy in one city. What percent of the days in that city were rainy?

Solve One-Step Equations

A statement that two expressions are equal is an equation. For example, $A = B$ is an equation that states that A is equal to B.

An equation is solved when a variable is replaced with a value that makes both sides of the equation equal. To make both sides equal the inverse operation is used. Addition and subtraction are inverses, and multiplication and division are inverses.

Example 1

Solve the equation $x - 10 = 35$.

Step 1 Find the solution by adding 10 to each side of the equation.

$$x - 10 = 35$$
$$x - 10 + 10 = 35 - 10$$
$$x = 45$$

Step 2 Check the solution.

$$x - 10 = 35$$
$$45 - 10 = 35$$
$$35 = 35$$

Both sides of the equation are equal, so $x = 45$.

Example 2

In the formula $a = bc$, find the value of c if $a = 20$ and $b = 2$.

Step 1 Rearrange the formula so the unknown value is by itself on one side of the equation by dividing both sides by b.

$$a = bc$$
$$\frac{a}{b} = \frac{bc}{b}$$
$$\frac{a}{b} = c$$

Step 2 Replace the variables a and b with the values that are given.

$$\frac{a}{b} = c$$
$$\frac{20}{2} = c$$
$$10 = c$$

Step 3 Check the solution.

$$a = bc$$
$$20 = 2 \times 10$$
$$20 = 20$$

Both sides of the equation are equal, so $c = 10$ is the solution when $a = 20$ and $b = 2$.

Practice Problem In the formula $h = gd$, find the value of d if $g = 12.3$ and $h = 17.4$.

SCIENCE SKILL HANDBOOK

MATH SKILL HANDBOOK

FOLDABLES HANDBOOK

REFERENCE HANDBOOK

GLOSSARY/ GLOSARIO

INDEX

Use Statistics

The branch of mathematics that deals with collecting, analyzing, and presenting data is statistics. In statistics, there are three common ways to summarize data with a single number—the mean, the median, and the mode.

The **mean** of a set of data is the arithmetic average. It is found by adding the numbers in the data set and dividing by the number of items in the set.

The **median** is the middle number in a set of data when the data are arranged in numerical order. If there were an even number of data points, the median would be the mean of the two middle numbers.

The **mode** of a set of data is the number or item that appears most often.

Another number that often is used to describe a set of data is the range. The **range** is the difference between the largest number and the smallest number in a set of data.

Example

The speeds (in m/s) for a race car during five different time trials are 39, 37, 44, 36, and 44.

To find the mean:

Step 1 Find the sum of the numbers.

$39 + 37 + 44 + 36 + 44 = 200$

Step 2 Divide the sum by the number of items, which is 5.

$200 \div 5 = 40$

The mean is 40 m/s.

To find the median:

Step 1 Arrange the measures from least to greatest.

36, 37, 39, 44, 44

Step 2 Determine the middle measure.

36, 37, <u>39</u>, 44, 44

The median is 39 m/s.

To find the mode:

Step 1 Group the numbers that are the same together.

44, 44, 36, 37, 39

Step 2 Determine the number that occurs most in the set.

<u>44, 44</u>, 36, 37, 39

The mode is 44 m/s.

To find the range:

Step 1 Arrange the measures from greatest to least.

44, 44, 39, 37, 36

Step 2 Determine the greatest and least measures in the set.

<u>44</u>, 44, 39, 37, 36

Step 3 Find the difference between the greatest and least measures.

$44 - 36 = 8$

The range is 8 m/s.

Practice Problem Find the mean, median, mode, and range for the data set 8, 4, 12, 8, 11, 14, 16.

A **frequency table** shows how many times each piece of data occurs, usually in a survey. **Table 1** below shows the results of a student survey on favorite color.

Table 1 Student Color Choice		
Color	**Tally**	**Frequency**
red	IIII	4
blue	ⅢⅠ	5
black	II	2
green	III	3
purple	ⅢⅠ II	7
yellow	ⅢⅠ I	6

Based on the frequency table data, which color is the favorite?

Use Geometry

The branch of mathematics that deals with the measurement, properties, and relationships of points, lines, angles, surfaces, and solids is called geometry.

Perimeter The **perimeter** (P) is the distance around a geometric figure. To find the perimeter of a rectangle, add the length and width and multiply that sum by two, or $2(l + w)$. To find perimeters of irregular figures, add the length of the sides.

Example 1

Find the perimeter of a rectangle that is 3 m long and 5 m wide.

Step 1 You know that the perimeter is 2 times the sum of the width and length.

$P = 2(3 \text{ m} + 5 \text{ m})$

Step 2 Find the sum of the width and length.

$P = 2(8 \text{ m})$

Step 3 Multiply by 2.

$P = 16 \text{ m}$

The perimeter is 16 m.

Example 2

Find the perimeter of a shape with sides measuring 2 cm, 5 cm, 6 cm, 3 cm.

Step 1 You know that the perimeter is the sum of all the sides.

$P = 2 + 5 + 6 + 3$

Step 2 Find the sum of the sides.

$P = 2 + 5 + 6 + 3$

$P = 16$

The perimeter is 16 cm.

Practice Problem Find the perimeter of a rectangle with a length of 18 m and a width of 7 m.

Practice Problem Find the perimeter of a triangle measuring 1.6 cm by 2.4 cm by 2.4 cm.

Area of a Rectangle The **area** (A) is the number of square units needed to cover a surface. To find the area of a rectangle, multiply the length times the width, or $l \times w$. When finding area, the units also are multiplied. Area is given in square units.

Example

Find the area of a rectangle with a length of 1 cm and a width of 10 cm.

Step 1 You know that the area is the length multiplied by the width.

$A = (1 \text{ cm} \times 10 \text{ cm})$

Step 2 Multiply the length by the width. Also multiply the units.

$A = 10 \text{ cm}^2$

The area is 10 cm².

Practice Problem Find the area of a square whose sides measure 4 m.

Area of a Triangle To find the area of a triangle, use the formula:

$A = \frac{1}{2}(\text{base} \times \text{height})$

The base of a triangle can be any of its sides. The height is the perpendicular distance from a base to the opposite endpoint, or vertex.

Example

Find the area of a triangle with a base of 18 m and a height of 7 m.

Step 1 You know that the area is $\frac{1}{2}$ the base times the height.

$A = \frac{1}{2}(18 \text{ m} \times 7 \text{ m})$

Step 2 Multiply $\frac{1}{2}$ by the product of 18×7. Multiply the units.

$A = \frac{1}{2}(126 \text{ m}^2)$

$A = 63 \text{ m}^2$

The area is 63 m².

Practice Problem Find the area of a triangle with a base of 27 cm and a height of 17 cm.

SCIENCE SKILL HANDBOOK

MATH SKILL HANDBOOK

FOLDABLES HANDBOOK

REFERENCE HANDBOOK

GLOSSARY/ GLOSARIO

INDEX

SCIENCE SKILL HANDBOOK

MATH SKILL HANDBOOK

FOLDABLES HANDBOOK

REFERENCE HANDBOOK

GLOSSARY/ GLOSARIO

INDEX

Circumference of a Circle The **diameter** (*d*) of a circle is the distance across the circle through its center, and the **radius** (r) is the distance from the center to any point on the circle. The radius is half of the diameter. The distance around the circle is called the **circumference** (C). The formula for finding the circumference is:

$$C = 2\pi r \text{ or } C = \pi d$$

The circumference divided by the diameter is always equal to 3.1415926… This nonterminating and nonrepeating number is represented by the Greek letter π (pi). An approximation often used for π is 3.14.

Example 1

Find the circumference of a circle with a radius of 3 m.

Step 1 You know the formula for the circumference is 2 times the radius times π.

$$C = 2\pi(3)$$

Step 2 Multiply 2 times the radius.

$$C = 6\pi$$

Step 3 Multiply by π.

$$C \approx 19 \text{ m}$$

The circumference is about 19 m.

Example 2

Find the circumference of a circle with a diameter of 24.0 cm.

Step 1 You know the formula for the circumference is the diameter times π.

$$C = \pi(24.0)$$

Step 2 Multiply the diameter by π.

$$C \approx 75.4 \text{ cm}$$

The circumference is about 75.4 cm.

Practice Problem Find the circumference of a circle with a radius of 19 cm.

Area of a Circle The formula for the area of a circle is: $A = \pi r^2$

Example 1

Find the area of a circle with a radius of 4.0 cm.

Step 1 $A = \pi(4.0)^2$

Step 2 Find the square of the radius.

$$A = 16\pi$$

Step 3 Multiply the square of the radius by π.

$$A \approx 50 \text{ cm}^2$$

The area of the circle is about 50 cm².

Example 2

Find the area of a circle with a radius of 225 m.

Step 1 $A = \pi(225)^2$

Step 2 Find the square of the radius.

$$A = 50625\pi$$

Step 3 Multiply the square of the radius by π.

$$A \approx 159043.1$$

The area of the circle is about 159043.1 m².

Example 3

Find the area of a circle whose diameter is 20.0 mm.

Step 1 Remember that the radius is half of the diameter.

$$A = \pi\left(\frac{20.0}{2}\right)^2$$

Step 2 Find the radius.

$$A = \pi(10.0)^2$$

Step 3 Find the square of the radius.

$$A = 100\pi$$

Step 4 Multiply the square of the radius by π.

$$A \approx 314 \text{ mm}^2$$

The area of the circle is about 314 mm².

Practice Problem Find the area of a circle with a radius of 16 m.

Volume The measure of space occupied by a solid is the **volume** (V). To find the volume of a rectangular solid multiply the length times width times height, or $V = l \times w \times h$. It is measured in cubic units, such as cubic centimeters (cm^3).

Example

Find the volume of a rectangular solid with a length of 2.0 m, a width of 4.0 m, and a height of 3.0 m.

Step 1 You know the formula for volume is the length times the width times the height.

$$V = 2.0 \text{ m} \times 4.0 \text{ m} \times 3.0 \text{ m}$$

Step 2 Multiply the length times the width times the height.

$$V = 24 \text{ m}^3$$

The volume is 24 m³.

Practice Problem Find the volume of a rectangular solid that is 8 m long, 4 m wide, and 4 m high.

To find the volume of other solids, multiply the area of the base times the height.

Example 1

Find the volume of a solid that has a triangular base with a length of 8.0 m and a height of 7.0 m. The height of the entire solid is 15.0 m.

Step 1 You know that the base is a triangle, and the area of a triangle is $\frac{1}{2}$ the base times the height, and the volume is the area of the base times the height.

$$V = \left[\frac{1}{2}(b \times h)\right] \times 15$$

Step 2 Find the area of the base.

$$V = \left[\frac{1}{2}(8 \times 7)\right] \times 15$$

$$V = \left(\frac{1}{2} \times 56\right) \times 15$$

Step 3 Multiply the area of the base by the height of the solid.

$$V = 28 \times 15$$

$$V = 420 \text{ m}^3$$

The volume is 420 m³.

Example 2

Find the volume of a cylinder that has a base with a radius of 12.0 cm, and a height of 21.0 cm.

Step 1 You know that the base is a circle, and the area of a circle is the square of the radius times π, and the volume is the area of the base times the height.

$$V = (\pi r^2) \times 21$$

$$V = (\pi 12^2) \times 21$$

Step 2 Find the area of the base.

$$V = 144\pi \times 21$$

$$V = 452 \times 21$$

Step 3 Multiply the area of the base by the height of the solid.

$$V \approx 9,500 \text{ cm}^3$$

The volume is about 9,500 cm³.

Example 3

Find the volume of a cylinder that has a diameter of 15 mm and a height of 4.8 mm.

Step 1 You know that the base is a circle with an area equal to the square of the radius times π. The radius is one-half the diameter. The volume is the area of the base times the height.

$$V = (\pi r^2) \times 4.8$$

$$V = \left[\pi\left(\frac{1}{2} \times 15\right)^2\right] \times 4.8$$

$$V = (\pi 7.5^2) \times 4.8$$

Step 2 Find the area of the base.

$$V = 56.25\pi \times 4.8$$

$$V \approx 176.71 \times 4.8$$

Step 3 Multiply the area of the base by the height of the solid.

$$V \approx 848.2$$

The volume is about 848.2 mm³.

Practice Problem Find the volume of a cylinder with a diameter of 7 cm in the base and a height of 16 cm.

SCIENCE SKILL HANDBOOK

MATH SKILL HANDBOOK

FOLDABLES HANDBOOK

REFERENCE HANDBOOK

GLOSSARY/ GLOSARIO

INDEX

Science Applications

SCIENCE SKILL HANDBOOK

MATH SKILL HANDBOOK

FOLDABLES HANDBOOK

REFERENCE HANDBOOK

GLOSSARY/ GLOSARIO

INDEX

Measure in SI

The metric system of measurement was developed in 1795. A modern form of the metric system, called the International System (SI), was adopted in 1960 and provides the standard measurements that all scientists around the world can understand.

The SI system is convenient because unit sizes vary by powers of 10. Prefixes are used to name units. Look at **Table 2** for some common SI prefixes and their meanings.

Table 2 Common SI Prefixes			
Prefix	**Symbol**	**Meaning**	
kilo–	k	1,000	thousandth
hecto–	h	100	hundred
deka–	da	10	ten
deci–	d	0.1	tenth
centi–	c	0.01	hundreth
milli–	m	0.001	thousandth

Example

How many grams equal one kilogram?

Step 1 Find the prefix *kilo–* in **Table 2.**

Step 2 Using **Table 2,** determine the meaning of *kilo–*. According to the table, it means 1,000. When the prefix *kilo–* is added to a unit, it means that there are 1,000 of the units in a "kilounit."

Step 3 Apply the prefix to the units in the question. The units in the question are grams. There are 1,000 grams in a kilogram.

Practice Problem Is a milligram larger or smaller than a gram? How many of the smaller units equal one larger unit? What fraction of the larger unit does one smaller unit represent?

Dimensional Analysis

Convert SI Units In science, quantities such as length, mass, and time sometimes are measured using different units. A process called dimensional analysis can be used to change one unit of measure to another. This process involves multiplying your starting quantity and units by one or more conversion factors. A conversion factor is a ratio equal to one and can be made from any two equal quantities with different units. If 1,000 mL equal 1 L then two ratios can be made.

$$\frac{1{,}000 \text{ mL}}{1 \text{ L}} = \frac{1 \text{ L}}{1{,}000 \text{ mL}} = 1$$

One can convert between units in the SI system by using the equivalents in **Table 2** to make conversion factors.

Example

How many cm are in 4 m?

Step 1 Write conversion factors for the units given. From **Table 2,** you know that 100 cm = 1 m. The conversion factors are

$$\frac{100 \text{ cm}}{1 \text{ m}} \text{ and } \frac{1 \text{ m}}{100 \text{ cm}}$$

Step 2 Decide which conversion factor to use. Select the factor that has the units you are converting from (m) in the denominator and the units you are converting to (cm) in the numerator.

$$\frac{100 \text{ cm}}{1 \text{ m}}$$

Step 3 Multiply the starting quantity and units by the conversion factor. Cancel the starting units with the units in the denominator. There are 400 cm in 4 m.

$$4 \cancel{\text{ m}} = \frac{100 \text{ cm}}{1 \cancel{\text{ m}}} = 400 \text{ cm}$$

Practice Problem How many milligrams are in one kilogram? (Hint: You will need to use two conversion factors from **Table 2.**)

Table 3 Unit System Equivalents

Type of Measurement	Equivalent
Length	1 in = 2.54 cm 1 yd = 0.91 m 1 mi = 1.61 km
Mass and weight*	1 oz = 28.35 g 1 lb = 0.45 kg 1 ton (short) = 0.91 tonnes (metric tons) 1 lb = 4.45 N
Volume	$1 \text{ in}^3 = 16.39 \text{ cm}^3$ 1 qt = 0.95 L 1 gal = 3.78 L
Area	$1 \text{ in}^2 = 6.45 \text{ cm}^2$ $1 \text{ yd}^2 = 0.83 \text{ m}^2$ $1 \text{ mi}^2 = 2.59 \text{ km}^2$ 1 acre = 0.40 hectares
Temperature	$°C = \dfrac{(°F - 32)}{1.8}$ $K = °C + 273$

*Weight is measured in standard Earth gravity.

Convert Between Unit Systems Table 3 gives a list of equivalents that can be used to convert between English and SI units.

Example

If a meterstick has a length of 100 cm, how long is the meterstick in inches?

Step 1 Write the conversion factors for the units given. From **Table 3,** 1 in = 2.54 cm.

$$\frac{1 \text{ in}}{2.54 \text{ cm}} \quad and \quad \frac{2.54 \text{ cm}}{1 \text{ in}}$$

Step 2 Determine which conversion factor to use. You are converting from cm to in. Use the conversion factor with cm on the bottom.

$$\frac{1 \text{ in}}{2.54 \text{ cm}}$$

Step 3 Multiply the starting quantity and units by the conversion factor. Cancel the starting units with the units in the denominator. Round your answer to the nearest tenth.

$$100 \text{ cm} \times \frac{1 \text{ in}}{2.54 \text{ cm}} = 39.37 \text{ in}$$

The meterstick is about 39.4 in long.

Practice Problem 1 A book has a mass of 5 lb. What is the mass of the book in kg?

Practice Problem 2 Use the equivalent for in and cm (1 in = 2.54 cm) to show how $1 \text{ in}^3 \approx 16.39 \text{ cm}^3$.

SCIENCE SKILL HANDBOOK

MATH SKILL HANDBOOK

FOLDABLES HANDBOOK

REFERENCE HANDBOOK

GLOSSARY/ GLOSARIO

INDEX

SCIENCE SKILL HANDBOOK

MATH SKILL HANDBOOK

FOLDABLES HANDBOOK

REFERENCE HANDBOOK

GLOSSARY/ GLOSARIO

INDEX

Precision and Significant Digits

When you make a measurement, the value you record depends on the precision of the measuring instrument. This precision is represented by the number of significant digits recorded in the measurement. When counting the number of significant digits, all digits are counted except zeros at the end of a number with no decimal point such as 2,050, and zeros at the beginning of a decimal such as 0.03020. When adding or subtracting numbers with different precision, round the answer to the smallest number of decimal places of any number in the sum or difference. When multiplying or dividing, the answer is rounded to the smallest number of significant digits of any number being multiplied or divided.

Example

The lengths 5.28 and 5.2 are measured in meters. Find the sum of these lengths and record your answer using the correct number of significant digits.

Step 1 Find the sum.

5.28 m	2 digits after the decimal
+ 5.2 m	1 digit after the decimal
10.48 m	

Step 2 Round to one digit after the decimal because the least number of digits after the decimal of the numbers being added is 1.

The sum is 10.5 m.

Practice Problem 1 How many significant digits are in the measurement 7,071,301 m? How many significant digits are in the measurement 0.003010 g?

Practice Problem 2 Multiply 5.28 and 5.2 using the rule for multiplying and dividing. Record the answer using the correct number of significant digits.

Scientific Notation

Many times numbers used in science are very small or very large. Because these numbers are difficult to work with scientists use scientific notation. To write numbers in scientific notation, move the decimal point until only one non-zero digit remains on the left. Then count the number of places you moved the decimal point and use that number as a power of ten. For example, the average distance from the Sun to Mars is 227,800,000,000 m. In scientific notation, this distance is 2.278×10^{11} m. Because you moved the decimal point to the left, the number is a positive power of ten.

The mass of an electron is about 0.000 000 000 000 000 000 000 000 000 000 911 kg. Expressed in scientific notation, this mass is 9.11×10^{-31} kg. Because the decimal point was moved to the right, the number is a negative power of ten.

Example

Earth is 149,600,000 km from the Sun. Express this in scientific notation.

Step 1 Move the decimal point until one non-zero digit remains on the left.

1.496 000 00

Step 2 Count the number of decimal places you have moved. In this case, eight.

Step 2 Show that number as a power of ten, 10^8.

Earth is 1.496×10^8 km from the Sun.

Practice Problem 1 How many significant digits are in 149,600,000 km? How many significant digits are in 1.496×10^8 km?

Practice Problem 2 Parts used in a high performance car must be measured to 7×10^{-6} m. Express this number as a decimal.

Practice Problem 3 A CD is spinning at 539 revolutions per minute. Express this number in scientific notation.

Make and Use Graphs

Data in tables can be displayed in a graph—a visual representation of data. Common graph types include line graphs, bar graphs, and circle graphs.

Line Graph A line graph shows a relationship between two variables that change continuously. The independent variable is changed and is plotted on the x-axis. The dependent variable is observed, and is plotted on the y-axis.

Table 4 Bicycle Race Data	
Time (h)	**Distance (km)**
0	0
1	8
2	16
3	24
4	32
5	40

Step 1 Determine the x-axis and y-axis variables. Time varies independently of distance and is plotted on the x-axis. Distance is dependent on time and is plotted on the y-axis.

Step 2 Determine the scale of each axis. The x-axis data ranges from 0 to 5. The y-axis data ranges from 0 to 50.

Step 3 Using graph paper, draw and label the axes. Include units in the labels.

Step 4 Draw a point at the intersection of the time value on the x-axis and corresponding distance value on the y-axis. Connect the points and label the graph with a title, as shown in **Figure 8.**

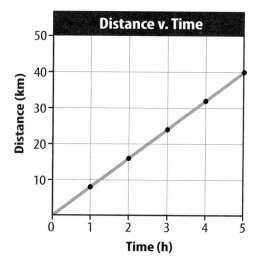

Figure 8 This line graph shows the relationship between distance and time during a bicycle ride.

Practice Problem A puppy's shoulder height is measured during the first year of her life. The following measurements were collected: (3 mo, 52 cm), (6 mo, 72 cm), (9 mo, 83 cm), (12 mo, 86 cm). Graph this data.

Find a Slope The slope of a straight line is the ratio of the vertical change, rise, to the horizontal change, run.

$$\text{Slope} = \frac{\text{vertical change (rise)}}{\text{horizontal change (run)}} = \frac{\text{change in } y}{\text{change in } x}$$

Example

Find the slope of the graph in **Figure 8**.

Step 1 You know that the slope is the change in y divided by the change in x.

$$\text{Slope} = \frac{\text{change in } y}{\text{change in } x}$$

Step 2 Determine the data points you will be using. For a straight line, choose the two sets of points that are the farthest apart.

$$\text{Slope} = \frac{(40 - 0) \text{ km}}{(5 - 0) \text{ h}}$$

Step 3 Find the change in y and x.

$$\text{Slope} = \frac{40 \text{ km}}{5 \text{ h}}$$

Step 4 Divide the change in y by the change in x.

$$\text{Slope} = \frac{8 \text{ km}}{\text{h}}$$

The slope of the graph is 8 km/h.

SCIENCE SKILL HANDBOOK

MATH SKILL HANDBOOK

FOLDABLES HANDBOOK

REFERENCE HANDBOOK

GLOSSARY/ GLOSARIO

INDEX

SCIENCE SKILL HANDBOOK

MATH SKILL HANDBOOK

FOLDABLES HANDBOOK

REFERENCE HANDBOOK

GLOSSARY/GLOSARIO

INDEX

Bar Graph To compare data that does not change continuously you might choose a bar graph. A bar graph uses bars to show the relationships between variables. The *x*-axis variable is divided into parts. The parts can be numbers such as years, or a category such as a type of animal. The *y*-axis is a number and increases continuously along the axis.

Example

A recycling center collects 4.0 kg of aluminum on Monday, 1.0 kg on Wednesday, and 2.0 kg on Friday. Create a bar graph of this data.

Step 1 Select the *x*-axis and *y*-axis variables. The measured numbers (the masses of aluminum) should be placed on the *y*-axis. The variable divided into parts (collection days) is placed on the *x*-axis.

Step 2 Create a graph grid like you would for a line graph. Include labels and units.

Step 3 For each measured number, draw a vertical bar above the *x*-axis value up to the *y*-axis value. For the first data point, draw a vertical bar above Monday up to 4.0 kg.

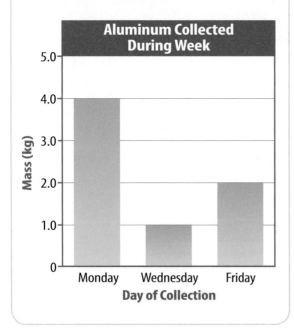

Practice Problem Draw a bar graph of the gases in air: 78% nitrogen, 21% oxygen, 1% other gases.

Circle Graph To display data as parts of a whole, you might use a circle graph. A circle graph is a circle divided into sections that represent the relative size of each piece of data. The entire circle represents 100%, half represents 50%, and so on.

Example

Air is made up of 78% nitrogen, 21% oxygen, and 1% other gases. Display the composition of air in a circle graph.

Step 1 Multiply each percent by 360° and divide by 100 to find the angle of each section in the circle.

$$78\% \times \frac{360°}{100} = 280.8°$$

$$21\% \times \frac{360°}{100} = 75.6°$$

$$1\% \times \frac{360°}{100} = 3.6°$$

Step 2 Use a compass to draw a circle and to mark the center of the circle. Draw a straight line from the center to the edge of the circle.

Step 3 Use a protractor and the angles you calculated to divide the circle into parts. Place the center of the protractor over the center of the circle and line the base of the protractor over the straight line.

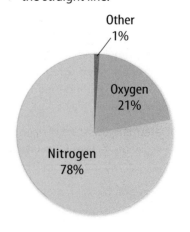

Practice Problem Draw a circle graph to represent the amount of aluminum collected during the week shown in the bar graph to the left.

Student Study Guides & Instructions
By Dinah Zike

1. You will find suggestions for Study Guides, also known as Foldables or books, in each chapter lesson and as a final project. Look at the end of the chapter to determine the project format and glue the Foldables in place as you progress through the chapter lessons.

2. Creating the Foldables or books is simple and easy to do by using copy paper, art paper, and internet printouts. Photocopies of maps, diagrams, or your own illustrations may also be used for some of the Foldables. Notebook paper is the most common source of material for study guides and 83% of all Foldables are created from it. When folded to make books, notebook paper Foldables easily fit into 11″ × 17″ or 12″ × 18″ chapter projects with space left over. Foldables made using photocopy paper are slightly larger and they fit into Projects, but snugly. Use the least amount of glue, tape, and staples needed to assemble the Foldables.

3. Seven of the Foldables can be made using either small or large paper. When 11″ × 17″ or 12″ × 18″ paper is used, these become projects for housing smaller Foldables. Project format boxes are located within the instructions to remind you of this option.

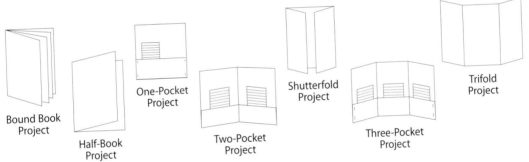

Bound Book Project

Half-Book Project

One-Pocket Project

Two-Pocket Project

Shutterfold Project

Three-Pocket Project

Trifold Project

4. Use one-gallon self-locking plastic bags to store your projects. Place strips of two-inch clear tape along the left, long side of the bag and punch holes through the taped edge. Cut the bottom corners off the bag so it will not hold air. Store this Project Portfolio inside a three-hole binder. To store a large collection of project bags, use a giant laundry-soap box. Holes can be punched in some of the Foldable Projects so they can be stored in a three-hole binder without using a plastic bag. Punch holes in the pocket books before gluing or stapling the pocket.

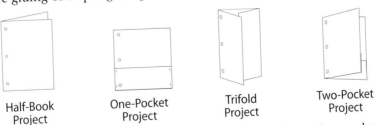

Half-Book Project

One-Pocket Project

Trifold Project

Two-Pocket Project

5. Maximize the use of the projects by collecting additional information and placing it on the back of the project and other unused spaces of the large Foldables.

SCIENCE SKILL HANDBOOK

MATH SKILL HANDBOOK

FOLDABLES HANDBOOK

REFERENCE HANDBOOK

GLOSSARY/GLOSARIO

INDEX

Half-Book Foldable® By Dinah Zike

Step 1 Fold a sheet of notebook or copy paper in half.

Label the exterior tab and use the inside space to write information.

PROJECT FORMAT
Use 11″ × 17″ or 12″ × 18″ paper on the horizontal axis to make a large project book.

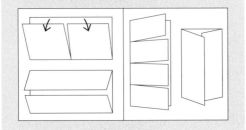

Variations
Paper can be folded horizontally, like a *hamburger* or vertically, like a *hot dog.*

A

B

C Half-books can be folded so that one side is ½ inch longer than the other side. A title or question can be written on the extended tab.

Worksheet Foldable or Folded Book® By Dinah Zike

Step 1 Make a half-book (see above) using work sheets, internet print-outs, diagrams, or maps.

Step 2 Fold it in half again.

Variations

A This folded sheet as a small book with two pages can be used for comparing and contrasting, cause and effect, or other skills.

B When the sheet of paper is open, the four sections can be used separately or used collectively to show sequences or steps.

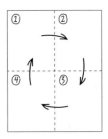

SCIENCE SKILL HANDBOOK

MATH SKILL HANDBOOK

FOLDABLES HANDBOOK

REFERENCE HANDBOOK

GLOSSARY/ GLOSARIO

INDEX

Two-Tab and Concept-Map Foldable® By Dinah Zike

Step 1 Fold a sheet of notebook or copy paper in half vertically or horizontally.

Step 2 Fold it in half again, as shown.

Step 3 Unfold once and cut along the fold line or valley of the top flap to make two flaps.

Variations

A Concept maps can be made by leaving a ½ inch tab at the top when folding the paper in half. Use arrows and labels to relate topics to the primary concept.

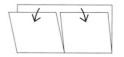

B Use two sheets of paper to make multiple page tab books. Glue or staple books together at the top fold.

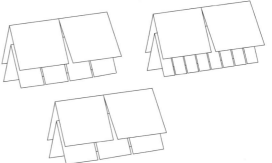

Three-Quarter Foldable® By Dinah Zike

Step 1 Make a two-tab book (see above) and cut the left tab off at the top of the fold line.

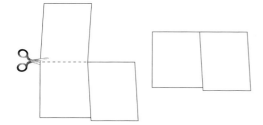

Variations

A Use this book to draw a diagram or a map on the exposed left tab. Write questions about the illustration on the top right tab and provide complete answers on the space under the tab.

B Compose a self-test using multiple choice answers for your questions. Include the correct answer with three wrong responses. The correct answers can be written on the back of the book or upside down on the bottom of the inside page.

SCIENCE SKILL HANDBOOK

MATH SKILL HANDBOOK

FOLDABLES HANDBOOK

REFERENCE HANDBOOK

GLOSSARY/ GLOSARIO

INDEX

Three-Tab Foldable® By Dinah Zike

Step 1 Fold a sheet of paper in half horizontally.

Step 2 Fold into thirds.

Step 3 Unfold and cut along the folds of the top flap to make three sections.

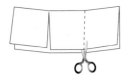

Variations

A Before cutting the three tabs draw a Venn diagram across the front of the book.

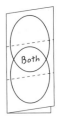

B Make a space to use for titles or concept maps by leaving a ½ inch tab at the top when folding the paper in half.

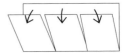

Four-Tab Foldable® By Dinah Zike

Step 1 Fold a sheet of paper in half horizontally.

Step 2 Fold in half and then fold each half as shown below.

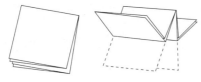

Step 3 Unfold and cut along the fold lines of the top flap to make four tabs.

Variations

A Make a space to use for titles or concept maps by leaving a ½ inch tab at the top when folding the paper in half.

B Use the book on the vertical axis, with or without an extended tab.

SCIENCE SKILL HANDBOOK

MATH SKILL HANDBOOK

FOLDABLES HANDBOOK

REFERENCE HANDBOOK

GLOSSARY/ GLOSARIO

INDEX

Folding Fifths for a Foldable® By Dinah Zike

Step 1 Fold a sheet of paper in half horizontally.

Step 2 Fold again so one-third of the paper is exposed and two-thirds are covered.

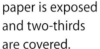

Step 3 Fold the two-thirds section in half.

Step 4 Fold the one-third section, a single thickness, backward to make a fold line.

Variations

A Unfold and cut along the fold lines to make five tabs.

B Make a five-tab book with a ½ inch tab at the top (see two-tab instructions).

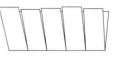

C Use 11″ × 17″ or 12″ × 18″ paper and fold into fifths for a five-column and/or row table or chart.

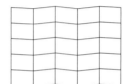

- -

Folded Table or Chart, and Trifold Foldable® By Dinah Zike

Step 1 Fold a sheet of paper in the required number of vertical columns for the table or chart.

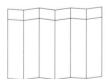

Step 2 Fold the horizontal rows needed for the table or chart.

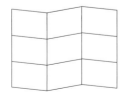

PROJECT FORMAT
Use 11″ × 17″ or 12″ × 18″ paper and fold it to make a large trifold project book or larger tables and charts.

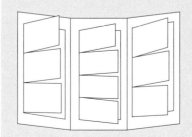

Variations

A Make a trifold by folding the paper into thirds vertically or horizontally.

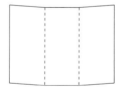

B Make a trifold book. Unfold it and draw a Venn diagram on the inside.

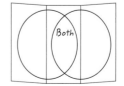

SCIENCE SKILL HANDBOOK

MATH SKILL HANDBOOK

FOLDABLES HANDBOOK

REFERENCE HANDBOOK

GLOSSARY/ GLOSARIO

INDEX

Two or Three-Pockets Foldable® By Dinah Zike

Step 1 Fold up the long side of a horizontal sheet of paper about 5 cm.

Step 2 Fold the paper in half.

Step 3 Open the paper and glue or staple the outer edges to make two compartments.

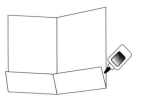

Variations

A Make a multi-page booklet by gluing several pocket books together.

B Make a three-pocket book by using a trifold (see previous instructions).

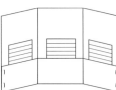

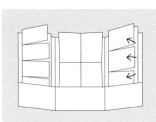

PROJECT FORMAT
Use 11″ × 17″ or 12″ × 18″ paper and fold it horizontally to make a large multi-pocket project.

Matchbook Foldable® By Dinah Zike

Step 1 Fold a sheet of paper almost in half and make the back edge about 1–2 cm longer than the front edge.

Step 2 Find the midpoint of the shorter flap.

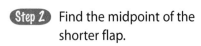

Step 3 Open the paper and cut the short side along the midpoint making two tabs.

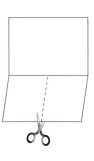

Step 4 Close the book and fold the tab over the short side.

Variations

A Make a single-tab matchbook by skipping Steps 2 and 3.

B Make two smaller matchbooks by cutting the single-tab matchbook in half.

SCIENCE SKILL HANDBOOK

MATH SKILL HANDBOOK

FOLDABLES HANDBOOK

REFERENCE HANDBOOK

GLOSSARY/ GLOSARIO

INDEX

Shutterfold Foldable® By Dinah Zike

Step 1 Begin as if you were folding a vertical sheet of paper in half, but instead of creasing the paper, pinch it to show the midpoint.

Step 2 Fold the top and bottom to the middle and crease the folds.

Variations

A Use the shutterfold on the horizontal axis.

B Create a center tab by leaving .5–2 cm between the flaps in Step 2.

PROJECT FORMAT
Use 11″ × 17″ or 12″ × 18″ paper and fold it to make a large shutterfold project.

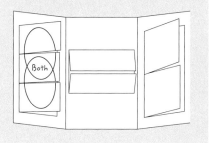

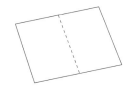

Four-Door Foldable® By Dinah Zike

Step 1 Make a shutterfold (see above).

Step 2 Fold the sheet of paper in half.

Step 3 Open the last fold and cut along the inside fold lines to make four tabs.

Variations

A Use the four-door book on the opposite axis.

B Create a center tab by leaving .5–2 cm between the flaps in Step 1.

SCIENCE SKILL HANDBOOK

MATH SKILL HANDBOOK

FOLDABLES HANDBOOK

REFERENCE HANDBOOK

GLOSSARY/ GLOSARIO

INDEX

Bound Book Foldable® By Dinah Zike

Step 1 Fold three sheets of paper in half. Place the papers in a stack, leaving about .5 cm between each top fold. Mark all three sheets about 3 cm from the outer edges.

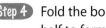
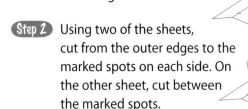

Step 2 Using two of the sheets, cut from the outer edges to the marked spots on each side. On the other sheet, cut between the marked spots.

Step 3 Take the two sheets from Step 1 and slide them through the cut in the third sheet to make a 12-page book.

Step 4 Fold the bound pages in half to form a book.

Variation

A Use two sheets of paper to make an eight-page book, or increase the number of pages by using more than three sheets.

PROJECT FORMAT

Use two or more sheets of 11″ × 17″ or 12″ × 18″ paper and fold it to make a large bound book project.

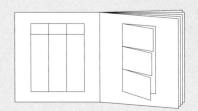

Accordian Foldable® By Dinah Zike

Step 1 Fold the selected paper in half vertically, like a *hamburger*.

Step 2 Cut each sheet of folded paper in half along the fold lines.

Step 3 Fold each half-sheet almost in half, leaving a 2 cm tab at the top.

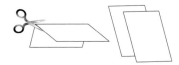

Step 4 Fold the top tab over the short side, then fold it in the opposite direction.

Variations

A Glue the straight edge of one paper inside the tab of another sheet. Leave a tab at the end of the book to add more pages.

B Tape the straight edge of one paper to the tab of another sheet, or just tape the straight edges of nonfolded paper end to end to make an accordian.

C Use whole sheets of paper to make a large accordian.

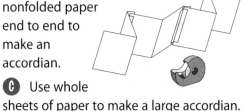

SCIENCE SKILL HANDBOOK

MATH SKILL HANDBOOK

FOLDABLES HANDBOOK

REFERENCE HANDBOOK

GLOSSARY/ GLOSARIO

INDEX

Layered Foldable® By Dinah Zike

Step 1 Stack two sheets of paper about 1–2 cm apart. Keep the right and left edges even.

Step 2 Fold up the bottom edges to form four tabs. Crease the fold to hold the tabs in place.

Step 3 Staple along the folded edge, or open and glue the papers together at the fold line.

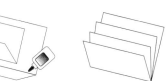

Variations

A Rotate the book so the fold is at the top or to the side.

B Extend the book by using more than two sheets of paper.

Envelope Foldable® By Dinah Zike

Step 1 Fold a sheet of paper into a *taco*. Cut off the tab at the top.

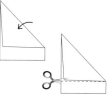

Step 2 Open the *taco* and fold it the opposite way making another *taco* and an X-fold pattern on the sheet of paper.

Step 3 Cut a map, illustration, or diagram to fit the inside of the envelope.

Step 4 Use the outside tabs for labels and inside tabs for writing information.

Variations

A Use 11" × 17" or 12" × 18" paper to make a large envelope.

B Cut off the points of the four tabs to make a window in the middle of the book.

SCIENCE SKILL HANDBOOK

MATH SKILL HANDBOOK

FOLDABLES HANDBOOK

REFERENCE HANDBOOK

GLOSSARY/ GLOSARIO

INDEX

Sentence Strip Foldable® By Dinah Zike

Step 1 Fold two sheets of paper in half vertically, like a *hamburger*.

Step 2 Unfold and cut along fold lines making four half sheets.

Step 3 Fold each half sheet in half horizontally, like a *hot dog*.

Step 4 Stack folded horizontal sheets evenly and staple together on the left side.

Step 5 Open the top flap of the first sentence strip and make a cut about 2 cm from the stapled edge to the fold line. This forms a flap that can be raisied and lowered. Repeat this step for each sentence strip.

Variations

A Expand this book by using more than two sheets of paper.

B Use whole sheets of paper to make large books.

Pyramid Foldable® By Dinah Zike

Step 1 Fold a sheet of paper into a *taco*. Crease the fold line, but do not cut it off.

Step 2 Open the folded sheet and refold it like a *taco* in the opposite direction to create an X-fold pattern.

Step 3 Cut one fold line as shown, stopping at the center of the X-fold to make a flap.

Step 4 Outline the fold lines of the X-fold. Label the three front sections and use the inside spaces for notes. Use the tab for the title.

Step 5 Glue the tab into a project book or notebook. Use the space under the pyramid for other information.

Step 6 To display the pyramid, fold the flap under and secure with a paper clip, if needed.

SCIENCE SKILL HANDBOOK

MATH SKILL HANDBOOK

FOLDABLES HANDBOOK

REFERENCE HANDBOOK

GLOSSARY/ GLOSARIO

INDEX

Single-Pocket or One-Pocket Foldable® By Dinah Zike

Step 1 Using a large piece of paper on a vertical axis, fold the bottom edge of the paper upwards, about 5 cm.

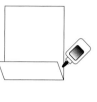

Step 2 Glue or staple the outer edges to make a large pocket.

PROJECT FORMAT
Use 11" × 17" or 12" × 18" paper and fold it vertically or horizontally to make a large pocket project.

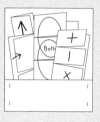

Variations

A Make the one-pocket project using the paper on the horizontal axis.

B To store materials securely inside, fold the top of the paper almost to the center, leaving about 2–4 cm between the paper edges. Slip the Foldables through the opening and under the top and bottom pockets.

Multi-Tab Foldable® By Dinah Zike

Step 1 Fold a sheet of notebook paper in half like a *hot dog*.

Step 2 Open the paper and on one side cut every third line. This makes ten tabs on wide ruled notebook paper and twelve tabs on college ruled.

Step 3 Label the tabs on the front side and use the inside space for definitions or other information.

Variation

A Make a tab for a title by folding the paper so the holes remain uncovered. This allows the notebook Foldable to be stored in a three-hole binder.

SCIENCE SKILL HANDBOOK

MATH SKILL HANDBOOK

FOLDABLES HANDBOOK

REFERENCE HANDBOOK

GLOSSARY/ GLOSARIO

INDEX

PERIODIC TABLE OF THE ELEMENTS

Element — Hydrogen
Atomic number — 1
Symbol — H
Atomic mass — 1.01
State of matter

🎈 Gas
💧 Liquid
⬜ Solid
⊙ Synthetic

1

Hydrogen
1
H 🎈
1.01

2

A column in the periodic table is called a **group.**

	1	2		3	4	5	6	7	8	9
1	Hydrogen 1 **H** 🎈 1.01									
2	Lithium 3 **Li** ⬜ 6.94	Beryllium 4 **Be** ⬜ 9.01								
3	Sodium 11 **Na** ⬜ 22.99	Magnesium 12 **Mg** ⬜ 24.31								
4	Potassium 19 **K** ⬜ 39.10	Calcium 20 **Ca** ⬜ 40.08	Scandium 21 **Sc** ⬜ 44.96	Titanium 22 **Ti** ⬜ 47.87	Vanadium 23 **V** ⬜ 50.94	Chromium 24 **Cr** ⬜ 52.00	Manganese 25 **Mn** ⬜ 54.94	Iron 26 **Fe** ⬜ 55.85	Cobalt 27 **Co** ⬜ 58.93
5	Rubidium 37 **Rb** ⬜ 85.47	Strontium 38 **Sr** ⬜ 87.62	Yttrium 39 **Y** ⬜ 88.91	Zirconium 40 **Zr** ⬜ 91.22	Niobium 41 **Nb** ⬜ 92.91	Molybdenum 42 **Mo** ⬜ 95.96	Technetium 43 **Tc** ⊙ (98)	Ruthenium 44 **Ru** ⬜ 101.07	Rhodium 45 **Rh** ⬜ 102.91
6	Cesium 55 **Cs** ⬜ 132.91	Barium 56 **Ba** ⬜ 137.33	Lanthanum 57 **La** ⬜ 138.91	Hafnium 72 **Hf** ⬜ 178.49	Tantalum 73 **Ta** ⬜ 180.95	Tungsten 74 **W** ⬜ 183.84	Rhenium 75 **Re** ⬜ 186.21	Osmium 76 **Os** ⬜ 190.23	Iridium 77 **Ir** ⬜ 192.22
7	Francium 87 **Fr** ⬜ (223)	Radium 88 **Ra** ⬜ (226)	Actinium 89 **Ac** ⬜ (227)	Rutherfordium 104 **Rf** ⊙ (267)	Dubnium 105 **Db** ⊙ (268)	Seaborgium 106 **Sg** ⊙ (271)	Bohrium 107 **Bh** ⊙ (272)	Hassium 108 **Hs** ⊙ (270)	Meitnerium 109 **Mt** ⊙ (276)

A row in the periodic table is called a **period.**

The number in parentheses is the mass number of the longest lived isotope for that element.

Lanthanide series	Cerium 58 **Ce** ⬜ 140.12	Praseodymium 59 **Pr** ⬜ 140.91	Neodymium 60 **Nd** ⬜ 144.24	Promethium 61 **Pm** ⊙ (145)	Samarium 62 **Sm** ⬜ 150.36	Europium 63 **Eu** ⬜ 151.96
Actinide series	Thorium 90 **Th** ⬜ 232.04	Protactinium 91 **Pa** ⬜ 231.04	Uranium 92 **U** ⬜ 238.03	Neptunium 93 **Np** ⊙ (237)	Plutonium 94 **Pu** ⊙ (244)	Americium 95 **Am** ⊙ (243)

Metal

Metalloid

Nonmetal

Recently discovered

				13	**14**	**15**	**16**	**17**	**18**
									Helium 2 He 4.00
				Boron 5 B 10.81	Carbon 6 C 12.01	Nitrogen 7 N 14.01	Oxygen 8 O 16.00	Fluorine 9 F 19.00	Neon 10 Ne 20.18
10	**11**	**12**		Aluminum 13 Al 26.98	Silicon 14 Si 28.09	Phosphorus 15 P 30.97	Sulfur 16 S 32.07	Chlorine 17 Cl 35.45	Argon 18 Ar 39.95
Nickel 28 Ni 58.69	Copper 29 Cu 63.55	Zinc 30 Zn 65.38		Gallium 31 Ga 69.72	Germanium 32 Ge 72.64	Arsenic 33 As 74.92	Selenium 34 Se 78.96	Bromine 35 Br 79.90	Krypton 36 Kr 83.80
Palladium 46 Pd 106.42	Silver 47 Ag 107.87	Cadmium 48 Cd 112.41		Indium 49 In 114.82	Tin 50 Sn 118.71	Antimony 51 Sb 121.76	Tellurium 52 Te 127.60	Iodine 53 I 126.90	Xenon 54 Xe 131.29
Platinum 78 Pt 195.08	Gold 79 Au 196.97	Mercury 80 Hg 200.59		Thallium 81 Tl 204.38	Lead 82 Pb 207.20	Bismuth 83 Bi 208.98	Polonium 84 Po (209)	Astatine 85 At (210)	Radon 86 Rn (222)
Darmstadtium 110 Ds (281)	Roentgenium 111 Rg (280)	Copernicium 112 Cn (285)		Ununtrium * 113 Uut (284)	Ununquadium * 114 Uuq (289)	Ununpentium * 115 Uup (288)	Ununhexium * 116 Uuh (293)		Ununoctium * 118 Uuo (294)

* The names and symbols for elements 113-116 and 118 are temporary. Final names will be selected when the elements' discoveries are verified.

Gadolinium 64 Gd 157.25	Terbium 65 Tb 158.93	Dysprosium 66 Dy 162.50	Holmium 67 Ho 164.93	Erbium 68 Er 167.26	Thulium 69 Tm 168.93	Ytterbium 70 Yb 173.05	Lutetium 71 Lu 174.97
Curium 96 Cm (247)	Berkelium 97 Bk (247)	Californium 98 Cf (251)	Einsteinium 99 Es (252)	Fermium 100 Fm (257)	Mendelevium 101 Md (258)	Nobelium 102 No (259)	Lawrencium 103 Lr (262)

SCIENCE SKILL HANDBOOK

MATH SKILL HANDBOOK

FOLDABLES HANDBOOK

REFERENCE HANDBOOK

GLOSSARY/ GLOSARIO

INDEX

SCIENCE SKILL HANDBOOK

MATH SKILL HANDBOOK

FOLDABLES HANDBOOK

REFERENCE HANDBOOK

GLOSSARY/ GLOSARIO

INDEX

Diversity of Life: Classification of Living Organisms

A six-kingdom system of classification of organisms is used today. Two kingdoms—Kingdom Archaebacteria and Kingdom Eubacteria—contain organisms that do not have a nucleus and that lack membrane-bound structures in the cytoplasm of their cells. The members of the other four kingdoms have a cell or cells that contain a nucleus and structures in the cytoplasm, some of which are surrounded by membranes. These kingdoms are Kingdom Protista, Kingdom Fungi, Kingdom Plantae, and Kingdom Animalia.

Kingdom Archaebacteria

one-celled; some absorb food from their surroundings; some are photosynthetic; some are chemosynthetic; many are found in extremely harsh environments including salt ponds, hot springs, swamps, and deep-sea hydrothermal vents

Kingdom Eubacteria

one-celled; most absorb food from their surroundings; some are photosynthetic; some are chemosynthetic; many are parasites; many are round, spiral, or rod-shaped; some form colonies

Kingdom Protista

Phylum Euglenophyta one-celled; photosynthetic or take in food; most have one flagellum; euglenoids

Kingdom Eubacteria
Bacillus anthracis

Phylum Chlorophyta
Desmids

Phylum Bacillariophyta one-celled; photosynthetic; have unique double shells made of silica; diatoms

Phylum Dinoflagellata one-celled; photosynthetic; contain red pigments; have two flagella; dinoflagellates

Phylum Chlorophyta one-celled, many-celled, or colonies; photosynthetic; contain chlorophyll; live on land, in freshwater, or salt water; green algae

Phylum Rhodophyta most are many-celled; photosynthetic; contain red pigments; most live in deep, saltwater environments; red algae

Phylum Phaeophyta most are many-celled; photosynthetic; contain brown pigments; most live in saltwater environments; brown algae

Phylum Rhizopoda one-celled; take in food; are free-living or parasitic; move by means of pseudopods; amoebas

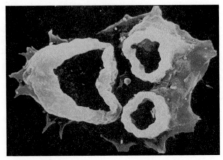

Amoeba

Phylum Zoomastigina one-celled; take in food; free-living or parasitic; have one or more flagella; zoomastigotes

Phylum Ciliophora one-celled; take in food; have large numbers of cilia; ciliates

Phylum Sporozoa one-celled; take in food; have no means of movement; are parasites in animals; sporozoans

Phylum Myxomycota
Slime mold

Phylum Oomycota
Phytophthora infestans

Phyla Myxomycota and Acrasiomycota one- or many-celled; absorb food; change form during life cycle; cellular and plasmodial slime molds

Phylum Oomycota many-celled; are either parasites or decomposers; live in freshwater or salt water; water molds, rusts and downy mildews

Kingdom Fungi

Phylum Zygomycota many-celled; absorb food; spores are produced in sporangia; zygote fungi; bread mold

Phylum Ascomycota one- and many-celled; absorb food; spores produced in asci; sac fungi; yeast

Phylum Basidiomycota many-celled; absorb food; spores produced in basidia; club fungi; mushrooms

Phylum Deuteromycota members with unknown reproductive structures; imperfect fungi; *Penicillium*

Phylum Mycophycota organisms formed by symbiotic relationship between an ascomycote or a basidiomycote and green alga or cyanobacterium; lichens

Lichens

SCIENCE SKILL HANDBOOK

MATH SKILL HANDBOOK

FOLDABLES HANDBOOK

REFERENCE HANDBOOK

GLOSSARY/ GLOSARIO

INDEX

Science Skill Handbook

Math Skill Handbook

Foldables Handbook

Reference Handbook

Glossary/Glosario

Index

Kingdom Plantae

Divisions Bryophyta (mosses), **Anthocerophyta** (hornworts), **Hepaticophyta** (liverworts), **Psilophyta** (whisk ferns) many-celled non-vascular plants; reproduce by spores produced in capsules; green; grow in moist, land environments

Division Lycophyta many-celled vascular plants; spores are produced in conelike structures; live on land; are photosynthetic; club mosses

Division Arthrophyta vascular plants; ribbed and jointed stems; scalelike leaves; spores produced in conelike structures; horsetails

Division Pterophyta vascular plants; leaves called fronds; spores produced in clusters of sporangia called sori; live on land or in water; ferns

Division Ginkgophyta deciduous trees; only one living species; have fan-shaped leaves with branching veins and fleshy cones with seeds; ginkgoes

Division Cycadophyta palmlike plants; have large, featherlike leaves; produces seeds in cones; cycads

Division Coniferophyta deciduous or evergreen; trees or shrubs; have needlelike or scalelike leaves; seeds produced in cones; conifers

Division Anthophyta
Tomato plant

Phylum Platyhelminthes
Flatworm

Division Gnetophyta shrubs or woody vines; seeds are produced in cones; division contains only three genera; gnetum

Division Anthophyta dominant group of plants; flowering plants; have fruits with seeds

Kingdom Animalia

Phylum Porifera aquatic organisms that lack true tissues and organs; are asymmetrical and sessile; sponges

Phylum Cnidaria radially symmetrical organisms; have a digestive cavity with one opening; most have tentacles armed with stinging cells; live in aquatic environments singly or in colonies; includes jellyfish, corals, hydra, and sea anemones

Phylum Platyhelminthes bilaterally symmetrical worms; have flattened bodies; digestive system has one opening; parasitic and free-living species; flatworms

Division Bryophyta
Liverwort

Phylum Chordata

Phylum Nematoda round, bilaterally symmetrical body; have digestive system with two openings; free-living forms and parasitic forms; roundworms

Phylum Mollusca soft-bodied animals, many with a hard shell and soft foot or footlike appendage; a mantle covers the soft body; aquatic and terrestrial species; includes clams, snails, squid, and octopuses

Phylum Annelida bilaterally symmetrical worms; have round, segmented bodies; terrestrial and aquatic species; includes earthworms, leeches, and marine polychaetes

Phylum Arthropoda largest animal group; have hard exoskeletons, segmented bodies, and pairs of jointed appendages; land and aquatic species; includes insects, crustaceans, and spiders

Phylum Echinodermata marine organisms; have spiny or leathery skin and a water-vascular system with tube feet; are radially symmetrical; includes sea stars, sand dollars, and sea urchins

Phylum Chordata organisms with internal skeletons and specialized body systems; most have paired appendages; all at some time have a notochord, nerve cord, gill slits, and a post-anal tail; include fish, amphibians, reptiles, birds, and mammals

SCIENCE SKILL HANDBOOK

MATH SKILL HANDBOOK

FOLDABLES HANDBOOK

REFERENCE HANDBOOK

GLOSSARY/ GLOSARIO

INDEX

Use and Care of a Microscope

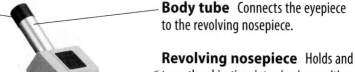

Eyepiece Contains magnifying lenses you look through.

Arm Supports the body tube.

Low-power objective Contains the lens with the lowest power magnification.

Stage clips Hold the microscope slide in place.

Coarse adjustment Focuses the image under low power.

Fine adjustment Sharpens the image under high magnification.

Body tube Connects the eyepiece to the revolving nosepiece.

Revolving nosepiece Holds and turns the objectives into viewing position.

High-power objective Contains the lens with the highest magnification.

Stage Supports the microscope slide.

Light source Provides light that passes upward through the diaphragm, the specimen, and the lenses.

Base Provides support for the microscope.

Caring for a Microscope

1. Always carry the microscope holding the arm with one hand and supporting the base with the other hand.

2. Don't touch the lenses with your fingers.

3. The coarse adjustment knob is used only when looking through the lowest-power objective lens. The fine adjustment knob is used when the high-power objective is in place.

4. Cover the microscope when you store it.

Using a Microscope

1. Place the microscope on a flat surface that is clear of objects. The arm should be toward you.

2. Look through the eyepiece. Adjust the diaphragm so light comes through the opening in the stage.

3. Place a slide on the stage so the specimen is in the field of view. Hold it firmly in place by using the stage clips.

4. Always focus with the coarse adjustment and the low-power objective lens first. After the object is in focus on low power, turn the nosepiece until the high-power objective is in place. Use ONLY the fine adjustment to focus with the high-power objective lens.

Making a Wet-Mount Slide

1. Carefully place the item you want to look at in the center of a clean, glass slide. Make sure the sample is thin enough for light to pass through.

2. Use a dropper to place one or two drops of water on the sample.

3. Hold a clean coverslip by the edges and place it at one edge of the water. Slowly lower the coverslip onto the water until it lies flat.

4. If you have too much water or a lot of air bubbles, touch the edge of a paper towel to the edge of the coverslip to draw off extra water and draw out unwanted air.

SCIENCE SKILL HANDBOOK

MATH SKILL HANDBOOK

FOLDABLES HANDBOOK

REFERENCE HANDBOOK

GLOSSARY/ GLOSARIO

INDEX

Glossary/Glosario

A science multilingual glossary is available on the science Web site. The glossary includes the following languages.

Arabic	Hmong	Tagalog
Bengali	Korean	Urdu
Chinese	Portuguese	Vietnamese
English	Russian	
Haitian Creole	Spanish	

Cómo usar el glosario en español:
1. Busca el término en inglés que desees encontrar.
2. El término en español, junto con la definición, se encuentran en la columna de la derecha.

Pronunciation Key
Use the following key to help you sound out words in the glossary.

a . . . . back (BAK)	**ew** . . . . foo**d** (FEWD)	
ay . . . . d**ay** (DAY)	**yoo** . . . . p**u**re (PYOOR)	
ah . . . . f**a**ther (FAH thur)	**yew** . . . . f**ew** (FYEW)	
ow . . . . fl**ow**er (FLOW ur)	**uh** . . . . comm**a** (CAH muh)	
ar . . . . c**ar** (CAR)	**u (+ con)** . . . . r**u**b (RUB)	
e . . . . l**e**ss (LES)	**sh** . . . . **sh**elf (SHELF)	
ee . . . . l**ea**f (LEEF)	**ch** . . . . na**t**ure (NAY chur)	
ih . . . . tr**i**p (TRIHP)	**g** . . . . **g**ift (GIHFT)	
i (i + com + e) **i**dea (i DEE uh)	**j** . . . . **g**em (JEM)	
oh . . . . g**o** (GOH)	**ing** . . . . s**ing** (SING)	
aw . . . . s**o**ft (SAWFT)	**zh** . . . . vi**si**on (VIH zhun)	
or . . . . **or**bit (OR buht)	**k** . . . . ca**k**e (KAYK)	
oy . . . . c**oi**n (COYN)	**s** . . . . **s**eed, **c**ent (SEED, SENT)	
oo . . . . f**oo**t (FOOT)	**z** . . . . **z**one, rai**s**e (ZOHN, RAYZ)	

English **A** **Español**

abiotic factor (ay bi AH tihk • FAK tuhr)/biosphere (BI uh sfihr) — factor abiótico/biosfera

abiotic factor (ay bi AH tihk • FAK tuhr): a nonliving thing in an ecosystem. (p. 707)

acid precipitation: precipitation that has a lower pH than that of normal rainwater (pH 5.6). (p. 825)

atmosphere (AT muh sfir): a thin layer of gases surrounding Earth. (p. 709)

factor abiótico: componente no vivo de un ecosistema. (pág. 707)

precipitación ácida: precipitación que tiene un pH más bajo que el del agua de la lluvia normal (pH 5.6). (pág. 825)

atmósfera: capa delgada de gases que rodean la Tierra. (pág. 709)

B

biome: a geographic area on Earth that contains ecosystems with similar biotic and abiotic features. (p. 777)

biosphere (BI uh sfihr): the parts of Earth and the surrounding atmosphere where there is life. (p. 741)

bioma: área geográfica en la Tierra que contiene ecosistemas con características bióticas y abióticas similares. (pág. 777)

biosfera: partes de la Tierra y de la atmósfera que la rodea donde hay vida. (pág. 741)

biotic factor (bi AH tihk • FAK tuhr): a living or once-living thing in an ecosystem. (p. 707)

biotic potential: the potential growth of a population if it could grow in perfect conditions with no limiting factors. (p. 744)

birthrate: the number of offspring produced by a population over a given time period. (p. 749)

factor biótico: vida cosa o anteriormente vida cosa en un ecosistema. (pág. 707)

potencial biótico: crecimiento potencial de una población si esta puede crecer en condiciones perfectas sin factores limitantes. (pág. 744)

tasa de nacimientos: número de crías que tiene una población durante un período de tiempo dado. (pág. 749)

C

carrying capacity: the largest number of individuals of one species that an ecosystem can support over time. (p. 745)

chemosynthesis (kee moh sihn THUH sus): the process during which producers use chemical energy in matter rather than light energy to make food. (p. 724)

climate: the long-term average weather conditions that occur in a particular region. (p. 708)

climax community: a stable community that no longer goes through major ecological changes. (p. 797)

commensalism: a symbiotic relationship that benefits one species but does not harm or benefit the other. (p. 764)

community: all the populations living in an ecosystem at the same time. (p. 742)

competition: the demand for resources, such as food, water, and shelter, in short supply in a community. (p. 743)

condensation (kahn den SAY shun): the process by which a gas changes to a liquid. (p. 714)

consumer: an organism that cannot make its own food and gets energy by eating other organisms. (p. 760)

coral reef: an underwater structure made from outside skeletons of tiny, soft-bodied animals called coral. (p. 793)

capacidad de carga: número mayor de individuos de una especie que un medioambiente puede mantener. (pág. 745)

quimiosíntesis: proceso durante el cual los productores usan la energía química en la materia en vez de la energía lumínica, para elaborar alimento. (pág. 724)

clima: promedio a largo plazo de las condiciones del tiempo atmosférico de una región en particular. (pág. 708)

comunidad clímax: comunidad estable que ya no sufrirá mayores cambios ecológicos. (pág. 797)

comensalismo: relación simbiótica que beneficia a una especie pero no causa daño ni beneficia a la otra. (pág. 764)

comunidad: todas las poblaciones que viven en un ecosistema, al mismo tiempo. (pág. 742)

competición: demanda de recursos, tales como alimento, agua y refugio, cuyo suministro es escaso en una comunidad. (pág. 743)

condensación: proceso mediante el cual un gas cambia a líquido. (pág. 714)

consumidor: organismo que no puede hacer sus propios alimentos y obtiene energía comiendo otros organismos. (pág. 760)

arrecife de coral: estructura bajo el agua formada por exoesqueletos de animales diminutos y de cuerpo blando. (pág. 793)

D

death rate: the number of individuals in a population that die over a given time period. (p. 749)

tasa de mortalidad: número de individuos que mueren en una población en un período de tiempo dado. (pág. 749)

desert: a biome that receives very little rain. (p. 778)

desierto: bioma que recibe muy poca lluvia. (pág. 778)

E

ecological succession: the process of one ecological community gradually changing into another. (p. 797)

ecosystem: all the living things and nonliving things in a given area. (p. 707)

endangered species: a species whose population is at risk of extinction. (p. 751)

energy pyramid: a model that shows the amount of energy available in each link of a food chain. (p. 728)

estuary (ES chuh wer ee): a coastal area where freshwater from rivers and streams mixes with salt water from seas or oceans. (p. 791)

eutrophication (yoo troh fuh KAY shun): the process of a body of water becoming nutrient-rich. (p. 800)

evaporation (ih va puh RAY shun): the process of a liquid changing to a gas at the surface of the liquid. (p. 714)

extinct species: a species that has died out and no individuals are left. (p. 751)

sucesión ecológica: proceso en el que una comunidad ecológica cambia gradualmente en otra. (pág. 797)

ecosistema: todos los seres vivos y los componentes no vivos de un área dada. (pág. 707)

especie en peligro: especie cuya población se encuentra en riesgo de extinción. (pág. 751)

pirámide energética: modelo que explica la cantidad de energía disponible en cada vínculo de una cadena alimentaria. (pág. 728)

estuario: zona costera donde el agua dulce de los ríos y arroyos se mezcla con el agua salada de los mares y los océanos. (pág. 791)

eutrofización: proceso por el cual un cuerpo de agua se vuelve rico en nutrientes. (pág. 800)

evaporación: proceso de cambio de un líquido a un gas en la superficie del líquido. (pág. 714)

especie extinta: especie que ha dejado de existir y no quedan individuos de ella. (pág. 751)

F

food chain: a model that shows how energy flows in an ecosystem through feeding relationships. (p. 726)

food web: a model of energy transfer that can show how the food chains in a community are interconnected. (p. 727)

cadena alimentaria: modelo que explica cómo la energía fluye en un ecosistema a través de relaciones alimentarias. (pág. 726)

red alimentaria: modelo de transferencia de energía que explica cómo las cadenas alimentarias están interconectadas en una comunidad. (pág. 727)

G

geothermal energy: thermal energy from Earth's interior. (p. 819)

global warming: an increase in the average temperature of Earth's surface. (p. 825)

grassland: a biome where grasses are the dominant plants. (p. 779)

energía geotérmica: energía térmica del interior de la Tierra. (pág. 819)

calentamiento global: incremento en la temperatura promedio de la superficie de la Tierra. (pág. 825)

pradera: bioma donde los pastos son las plantas dominantes. (pág. 779)

H

habitat: the place within an ecosystem where an organism lives; provides the biotic and abiotic factors an organism needs to survive and reproduce. (p. 759)

hábitat: lugar en un ecosistema donde vive un organismo; proporciona los factores bióticos y abióticos de un organismo necesita para sobrevivir y reproducirse. (pág. 759)

I

inexhaustible resource: a natural resource that will not run out, no matter how much of it people use. (p. 818)

intertidal zone: the ocean shore between the lowest low tide and the highest high tide. (p. 793)

recurso inagotable: recurso natural que no se acabará, sin importar cuánto lo usen las personas. (pág. 818)

zona intermareal: playa en medio de la marea baja más baja y la marea alta más alta. (pág. 793)

L

limiting factor: a factor that can limit the growth of a population. (p. 743)

factor limitante: factor que puede limitar el crecimiento de una población. (pág. 743)

M

migration: the instinctive, seasonal movement of a population of organisms from one place to another. (p. 752)

mutualism: a symbiotic relationship in which both organisms benefit. (p. 763)

migración: movimiento instintivo de temporada de una población de organismos de un lugar a otro. (pág. 752)

mutualismo: relación simbiótica en la cual los dos organismos se benefician. (pág. 763)

N

natural resource: part of the environment that supplies material useful or necessary for the survival of living things. (p. 813)

niche (NICH): the way a species interacts with abiotic and biotic factors to obtain food, find shelter, and fulfill other needs. (p. 759)

nitrogen fixation (NI truh jun • fihk SAY shun): the process that changes atmospheric nitrogen into nitrogen compounds that are usable by living things. (p. 716)

nonrenewable resource: a natural resource that is used up faster than it can be replaced by natural processes. (p. 814)

recurso natural: parte del medioambiente que suministra material útil o necesario para que los seres vivos sobrevivan. (pág. 813)

nicho: forma de una especie interacciona con los factores abióticos y bióticos para obtener comida, encontrar refugio, y satisfacer otras necesidades. (pág. 759)

fijación del nitrógeno: proceso que cambia el nitrógeno atmosférico en componentes de nitrógeno útiles para los seres vivos. (pág. 716)

recurso no renovable: recurso natural que se usa más rápidamente de lo que se puede reemplazar por procesos naturales. (pág. 814)

SCIENCE SKILL HANDBOOK

MATH SKILL HANDBOOK

REFERENCE HANDBOOK

GLOSSARY/ GLOSARIO

INDEX

O

ozone layer: the layer of atmosphere that prevents most harmful ultraviolet (UV) radiation from reaching Earth. (p. 824)

capa de ozono: capa de la atmósfera que evita que la mayor parte de la radiación ultravioleta dañina llegue a la Tierra. (pág. 824)

P

parasitism: a symbiotic relationship in which one organism benefits and the other is harmed. (p. 764)

photochemical smog: air pollution that forms from the interaction between chemicals in the air and sunlight. (p. 824)

photosynthesis (foh toh SIHN thuh sus): a series of chemical reactions that convert light energy, water, and carbon dioxide into the food-energy molecule glucose and give off oxygen. (p. 724)

pioneer species: the first species that colonizes new or undisturbed land. (p. 798)

pollution: the contamination of the environment with substances that are harmful to life. (p. 823)

population density: the size of a population compared to the amount of space available. (p. 744)

population: all the organisms of the same species that live in the same area at the same time. (p. 742)

precipitation (prih sih puh TAY shun): water, in liquid or solid form, that falls from the atmosphere. (p. 714)

producer: an organism that uses an outside energy source, such as the Sun, and produces its own food. (p. 760)

parasitismo: relación simbiótica en la cual se perjudica organismo se beneficia y el otro. (pág. 764)

smog fotoquímico: polución del aire que se forma de la interacción entre los químicos en el aire y la luz solar. (pág. 824)

fotosíntesis: serie de reacciones químicas que convierten la energía lumínica, el agua y el dióxido de carbono en glucosa, una molécula de energía alimentaria, y libera oxígeno. (pág. 724)

especie pionera: primera especie que coloniza tierra nueva o tierra virgen. (pág. 798)

polución: contaminación del medioambiente con sustancias dañinas para la vida. (pág. 823)

densidad poblacional: tamaño de una población comparado con la cantidad de espacio disponible. (pág. 744)

población: todos los organismos de la misma especie que viven en la misma área al mismo tiempo. (pág. 742)

precipitación: agua, en forma líquida o sólida, que cae de la atmósfera. (pág. 714)

productor: organismo que utiliza una fuente de energía exterior, como el Sol, y produce sus propios alimentos. (pág. 760)

R

recycling: manufacturing new products out of used products. (p. 836)

renewable resource: a natural resource that can be replenished by natural processes at least as quickly as it is used. (p. 816)

reciclaje: fabricación de productos nuevos hechos de productos usados. (pág. 836)

recurso renovable: recurso natural que se reabastece mediante procesos naturales tan rápidamente como se usa. (pág. 816)

S

salinity (say LIH nuh tee): a measure of the mass of dissolved salts in a mass of water. (p. 787)

sustainability: meeting human needs in ways that ensure future generations also will be able to meet their needs. (p. 835)

symbiosis (sihm bee OH sus): a close, long-term relationship between two species that usually involves an exchange of food or energy. (p. 763)

salinidad: medida de la masa de sales disueltas en una masa de agua. (pág. 787)

sostenibilidad: satisfacción de las necesidades humanas de forma que se asegure que las generaciones futuras también podrán satisfacer sus necesidades. (pág. 835)

simbiosis: relación intrínseca a largo plazo entre dos especies que generalmente involucra intercambio de alimento o energía. (pág. 763)

T

taiga (TI guh): a forest biome consisting mostly of cone-bearing evergreen trees. (p. 783)

temperate: the term describing any region of Earth between the tropics and the polar circles. (p. 781)

threatened species: a species at risk, but not yet endangered. (p. 751)

tundra (TUN druh): a biome that is cold, dry, and treeless. (p. 783)

taiga: bioma de bosque constituido en su mayoría por coníferas perennes. (pág. 783)

temperatura: término que describe cualquier región de la Tierra entre los trópicos y los círculos polares. (pág. 781)

especie amenazada: especie en riesgo, pero que todavía no está en peligro. (pág. 751)

tundra: bioma frío, seco y sin árboles. (pág. 783)

W

wetland: an aquatic ecosystem that has a thin layer of water covering soil that is wet most of the time. (p. 790)

humedal: ecosistema acuático que tiene una capa delgada de suelo cubierto de agua que permanece húmedo la mayor parte del tiempo. (pág. 790)

SCIENCE SKILL HANDBOOK

MATH SKILL HANDBOOK

REFERENCE HANDBOOK

GLOSSARY/ GLOSARIO

INDEX

Index

Italic numbers = illustration/photo   **Bold numbers** = vocabulary term
lab = indicates entry is used in a lab on this page

A

Abiotic factor(s)
in aquatic ecosystems, 787
effects of, 711
explanation of, **707,** *708, 710*
interaction between animals and, *709*
plant growth and, 730–731 *lab*
types of, 708–709, *732*
Academic Vocabulary, 719, 753, 825.
See also **Vocabulary**
Acid precipitation, 825
Agriculture
pollution from, 827, *827*
Air
as renewable resource, 816
Air pollution. *See also* **Pollution**
explanation of, 824, 832 *lab*
sources of, *824,* 824–825, *825*
Alternative fuel(s), 833
Aquatic ecosystem(s)
estuary, 791, *791*
explanation of, 787
freshwater, 787 *lab, 788,* 788–789, *789*
ocean, 787 *lab, 792,* 792–793, *793*
wetland, 790, *790*
Aquatic succession, 800, *800*
Arengo, Felicity, 747
Atmosphere
effect in ecosystem, 709
explanation of, **709**
gases in, 709, *709,* 715, *715,* 718, 719
on Mars, 711
Automobile(s)
new technologies for, 834, *834*

B

Bacteria
breakdown of dead organisms by,
716, *716,* 736
chemosynthetic, 724
explanation of, **717**
in nitrogen cycle, 715–716
Big Idea, 704, 732, 738, 768, 774, 804
Review, 735, 771, 807
Biodiesel, 833
Biodiversity
in deserts, 778
in estuaries, 791
in freshwater ecosystems, 788, 789
in grasslands, 779
in oceans, 792, 793
in taiga, 783
in temperate deciduous forests, 782
in temperate rain forests, 781

in tropical rain forests, 780
in tundra, 783
in wetlands, 790
Biome(s). *See also* **Aquatic ecosys-**
tems; Ecosystems; Land biomes
explanation of, *777,* **777**
modeling of, 802–803 *lab*
Biosphere
ecological systems and, 741
example of, 742, *742*
explanation of, **741**
Biotic factor(s), 707, *710*
Biotic potential, 744
Birthrate
explanation of, **749**
population size and, 757 *lab*
Boxer crab(s), 763, *763*
Brumbaugh, Dan, 795

C

Calcium
as mineral, 815
Carbon
in air, 719
in organisms, 718
in soil, 718
Carbon cycle
explanation of, *718,* 718–719, 724
interaction between oxygen cycle
and, 717
organisms and, 718
Carbon dioxide (CO₂)
in atmosphere, 711, 718, 719
climate change and, 825
greenhouse effect and, 719, *719*
production of, 717
Careers in Science, 747
Carnivore(s)
explanation of, 725
method of obtaining energy by, 760
Carrying capacity
of Earth, 753
explanation of, **745**
Cellular respiration, 714
Chapter Review, 734–735, 770–771,
806–807, 842–843
Chemical(s)
pollution from, 823
Chemosynthesis
energy and, 724
explanation of, **724**
Chemosynthetic bacteria, 724
Chlorofluorocarbon(s) (CFC)
effects of, 711
explanation of, 824, *833*
phasing out of, 833

Chlorophyll
distribution of, 741, *741*
Clean Air Act, 832
Climate
in China, 777 *lab*
effect in ecosystem, 708
explanation of, **708**
Climate change. *See also* **Global**
warming
causes of, 825
Climax community, 797
Cloud(s)
formation of, 714
Coal
as fossil fuel, 814, *814*
pollution from burning, 825
Coal ash sludge, 827
Coal mining
pollution from, 827
Coastal ocean(s), 793
Commensalism, 764, *764*
Common Use. *See* **Science Use v.**
Common Use
Community(ies)
cooperative relationships in, 763, *763*
energy in, 760
explanation of, **742,** 759, 768, **797**
predator-prey relationships in, 762
symbiotic relationships in, *763,*
763–764, *764*
Compact fluorescent lightbulb(s)
(CFL), 832
Competition, 743
Compost, 836
Condensation
explanation of, **714**
in water cycle, 714
Consumer(s)
energy and, 725–727
explanation of, 725, **760**
Cooperative relationship(s), 763, *763*
Copper, 815
Coral reef(s), 793
Critical thinking, 710, 720, 735, 746,
756, 765, 771, 784, 794, 801, 807
Cycles of matter. *See* **Carbon cycle;**
Nitrogen cycle; Oxygen cycle;
Water cycle

D

Death rate
explanation of, **749**
population size and, 757 *lab*
Decomposer(s), 725

SCIENCE SKILL HANDBOOK

MATH SKILL HANDBOOK

REFERENCE HANDBOOK

GLOSSARY/ GLOSARIO

INDEX

Credits

Photo Credits

Front Cover Spine Photodisc/Getty Images; Back Cover Thinkstock/Getty Images; Inside front,back cover Thinkstock/Getty Images; Connect Ed (t) Richard Hutchings, (c)Getty Images, (b)Jupiterimages/ThinkStock/Alamy; i ThinkStock/Getty Images; viii–ix The McGraw-Hill Companies; ix (b)Fancy Photography/Veer; 702 (t to b)Georg Gerster/Photo Researchers, Inc., (2) Keren Su/Getty Images, (3)Smneedham/Getty Images, (4)Courtesy of the University of Central Florida; 703 (t)SuperStock/SuperStock, (b)Stuart Fox/ Getty Images; 704–705 MICHAEL S. QUINTON/National Geographic Image Collection; 706 (c)George H.H. Huey; 708–709 Gerry Ellis/Minden Pictures/ Getty Images; 709 Nigel Cattlin/Alamy; 710 (t)MICHAEL S. QUINTON/ National Geographic Image Collection, (b)Gerry Ellis/Minden Pictures/Getty Images; 711 NASA-JPL; 712 IIC/Axiom/Gatty Images; 713 Hutchings Photography/Digital Light Source; 715 Horizons Companies; 716 (t)Visuals Unlimited/CORBIS, (b) Melbourne Etc/Alamy; 720 Melbourne Etc/Alamy; 721 (all)Hutchings Photography/Digital Light Source; 722 (c)Art Wolfe/ Getty Images; 723 Hutchings Photography/Digital Light Source; 724 NOAA; 725 (l)Digital Vision/PunchStock, (cl) SA Team/Foto Natura/Minden Pictures, (c)BananaStock/PunchStock, (cr)Colin Young-Wolff/PhotoEdit, (r)Mark Steinmetz; 730 (6)Macmillan/McGraw-Hill, (others)Hutchings Photography/Digital Light Source; 732 George H.H. Huey; 735 MICHAEL S. QUINTON/National Geographic Image Collection; 738–739 Heidi & Hans-Jurgen Koch/Minden Pictures; 740 Martin Harvey/CORBIS; 741 (t)Hutchings Photography/Digital Light Source, (b)NASA Goddard Space Flight Center (NASA-GSFC); 742 (l)NASA, (r)Nigel J. Dennis/Gallo Images/CORBIS; 743 Hutchings Photography/Digital Light Source; 744 (t,b)Raymond Gehman/ CORBIS; 745 (t)Riaan Janse van Rensburg/Alamy, (b)Gerry Ellis/Minden Pictures/Getty Images; 746 (t)NASA Goddard Space Flight Center (NASA-GSFC), (b)Nigel J. Dennis/Gallo Images/CORBIS; 747 (t)Andoni Canela/ Photolibrary, (c)O. Rocha/American Museum of Natural History, (b)Pete Oxford/Minden Pictures; 748 Dr Jeremy Burgess/Photo Researchers; 749 (t) Hutchings Photography/Digital Light Source, (b)TomVezo.com; 750 Dr. David Phillips/Visuals Unlimited/Getty Images; 751 (t)Mary Evans Picture Library/Alamy, (c)Paul Souders/CORBIS, (b)Tom Brakefield/Getty Images; 752 Hutchings Photography/Digital Light Source; 753 Ricardo Beliel/ BrazilPhotos; 754 Blend Images/Jupiterimages; 755 BEAWIHARTA/Reuters/ CORBIS; 756 (t)TomVezo.com, (c)Mary Evans Picture Library/Alamy, (b)Tom Brakefield/Getty Images; 757 Tom & Pat Leeson; 758 Duncan Usher/Minden Pictures; 759 Jacques Jangoux/Photo Researchers; 760 (t)Hutchings Photography/Digital Light Source; 762 Bach/zefa/CORBIS; 763 (t)CORBIS, (b)Mark Strickland/SeaPics.com; 764 (t)Carol & Don Spencer/Visual Unlimited/Getty Images, (b)Michael & Patricia Fogden/Minden Pictures/ Getty Images; 765 (t)Jacques Jangoux/Photo Researchers, (c)Bach/zefa/ CORBIS, (b)Michael & Patricia Fogden/Minden Pictures/Getty Images; 766 (l,br)Hutchings Photography/Digital Light Source, (tr)Digital Vision/Getty Images; 768 (t)Nigel J. Dennis/Gallo Images/CORBIS, (c)Dr Jeremy Burgess/ Photo Researchers, (b)CORBIS; 771 Heidi & Hans-Jurgen Koch/Minden Pictures; 774–775 Roger Ressmeyer/CORBIS; 776 Frans Lanting/CORBIS; 778 (t)Tom Vezo/Minden Pictures/Getty Images, (c)Hutchings Photography/ Digital Light Source, (b)David Muench; 779 (t)Laura Romin & Larry Dalton/ Alamy, (tc)Jim Brandenburg/Minden Pictures, (b)Chuck Haney/ DanitaDelimont.com, (bc)Comstock/PunchStock; 780 (tl)age fotostock/ Photolibrary, (tr)Gavriel Jecan/Getty Images, (bl)Jacques Jangoux/Mira. com, (br)Frans Lanting/CORBIS; 781 (t)M DeFreitas/Getty Images, (bl)Doug Sherman/Geofile, (br)Comstock Images/Alamy; 782 (t)Tom Till, (bl)Bruce Lichtenberger/Peter Arnold, Inc., (br)Russ Munn/CORBIS; 783 (t)age fotostock/SuperStock, (tc)Marvin Dembinsky Photo Associates/Alamy, (b) Andre Gallant/Getty Images, (bc)age fotostock/SuperStock; 784 (t)Frans Lanting/CORBIS, (b)Tom Till; 785 (b)Andre Gallant/Getty Images, (bcr)Andre Gallant/Getty Images; 786 Woodfall/Photoshot; 787 (tr)Hutchings Photography/Digital Light Source; 788 (tl)Comstock/PunchStock, (tr)Arthur Morris/Visuals Unlimited/Getty Images, (cl)Paul Nicklen/NGS/Getty Images, (cr)Stephen Dalton/Minden Pictures, (b)Medioimages/PunchStock, (bcl)Dr. Marli Mill/Visuals Unlimited/Getty Images; 789 (t)Jean-Paul Ferrero/ Minden Picture, (tc)Lynn & Donna Rogers/Peter Arnold, Inc., (b)Image Ideas/PictureQuest, (bc)Photograph by Tim McCabe, courtesy USDA Natural Resources Conservation Service; 790 (t)Michael S. Quinton/National Geographic/Getty Images, (c)James L. Amos/Peter Arnold, Inc., (b)Steve Bly/Getty Images; 791 (t)Tom & Therisa Stack/Tom Stack & Associates, (bl) B. Moose Peterson, (br)David Noton Photography/Alamy; 792 (t)Camille Lusardi/Photolibrary, (c)Gregory Ochocki/Photo Researchers, (b)Doug Allan/ Getty Images; 793 (t)Gavriel Jecan/Photolibrary, (b)Comstock Images/ PictureQuest, (br)Hutchings Photography/Digital Light Source; 794 (t)Jean-Paul Ferrero/Minden Picture, (c)Steve Bly/Getty Images, (b)Comstock Images/PictureQuest; 795 (t)K. Holmes/American Museum of Natural History, (c)K. Frey/American Museum of Natural History, (b)Stephen Frink/ Getty Images, (bkgd)Wayne Levin/Getty Images; 796 Paul Bradforth/ Alamy; 802 (others)Hutchings Photography/Digital Light Source, (2,3,4) Macmillan/McGraw-Hill; 803 (tr)Hutchings Photography/Digital Light Source; 804 (t)David Muench, (c)Tom & Therisa Stack/Tom Stack & Associates; 807 Roger Ressmeyer/CORBIS; 808 (6)Macmillan/McGraw-Hill; 810–811 Sarah Leen/National Geographic/Getty Images; 812 Ron Chapple Stock/Alamy; 813 Creatas/PunchStock; 814 CHRIS JAMES/Peter Arnold, Inc.; 815 (l)Mark Steinmetz, (r)ImageState/age fotostock; 816 Gary Braasch/ Getty Images; 819 inga spence/Alamy; 820 (t)Creatas/PunchStock, (b)Ron Chapple Stock/Alamy; 821 PhotoLink/Getty Images; 822 Thomas R. Fletcher/Alamy; 823 (t)Hutchings Photography/Digital Light Source, (b) NATALIE B. FOBES/National Geographic Stock; 825 Robert Jureit/Getty Images; 826 Hutchings Photography/Digital Light Source; 827 (l)Scott Bauer/USDA, (r)Harrison Shull/Getty Images; 828 (t)NATALIE B. FOBES/ National Geographic Stock, (c)Robert Jureit/Getty Images, (b)Harrison Shull/ Getty Images; 829 (t)The Mcgraw-Hill Companies, (c)Rob Melnychuk/Brand X/CORBIS, (b)Hutchings Photography/Digital Light Source; 830 NASA Jet Propulsion Laboratory (NASA-JPL); 831 Butch Martin/Getty Images; 832 (t) JAMES L. STANFIELD/National Geographic Stock, (b)Mark Steinmetz; 835 (l) Jonathan Nourok/PhotoEdit, (c)Steve Skjold/Alamy Images, (r)David Young-Wolff/PhotoEdit; 836 (l)Peter Starman/Getty Images, (r)Tony Craddock/ Photo Researchers; 837 (t)NASA Jet Propulsion Laboratory (NASA-JPL), (b) Tony Craddock/Photo Researchers; 839 Mark Steinmetz; 840 (t)Mark Steinmetz, (c)Thomas R. Fletcher/Alamy, (b)JAMES L. STANFIELD/National Geographic Stock; 843 Sarah Leen/National Geographic/Getty Images; SR-00–SR-01 (bkgd)Gallo Images-Neil Overy/Getty Images; SR-02 Hutchings Photography/Digital Light Source; SR-06 Michell D. Bridwell/ PhotoEdit; SR-07 (t)The McGraw-Hill Companies, (b)Dominic Oldershaw; SR-08 StudiOhio; SR-09 Timothy Fuller; SR-10 Aaron Haupt; SR-42 (c) NIBSC/Photo Researchers, Inc., (r)Science VU/Drs. D.T. John & T.B. Cole/ Visuals Unlimited, Inc.; Stephen Durr; SR-43 (t)Mark Steinmetz, (r)Andrew Syred/Science Photo Library/Photo Researchers, (br)Rich Brommer; SR-44 (l)Lynn Keddie/Photolibrary, (tr)G.R. Roberts; David Fleetham/Visuals Unlimited/Getty Images; SR-45 Gallo Images/CORBIS; SR-46 Matt Meadows.

PERIODIC TABLE OF THE ELEMENTS

Element — Hydrogen
Atomic number — 1
Symbol — H
Atomic mass — 1.01
State of matter

Gas
Liquid
Solid
Synthetic

A column in the periodic table is called a **group.**

A row in the periodic table is called a **period.**

	1	2	3	4	5	6	7	8	9
1	Hydrogen 1 H 1.01								
2	Lithium 3 Li 6.94	Beryllium 4 Be 9.01							
3	Sodium 11 Na 22.99	Magnesium 12 Mg 24.31							
4	Potassium 19 K 39.10	Calcium 20 Ca 40.08	Scandium 21 Sc 44.96	Titanium 22 Ti 47.87	Vanadium 23 V 50.94	Chromium 24 Cr 52.00	Manganese 25 Mn 54.94	Iron 26 Fe 55.85	Cobalt 27 Co 58.93
5	Rubidium 37 Rb 85.47	Strontium 38 Sr 87.62	Yttrium 39 Y 88.91	Zirconium 40 Zr 91.22	Niobium 41 Nb 92.91	Molybdenum 42 Mo 95.96	Technetium 43 Tc (98)	Ruthenium 44 Ru 101.07	Rhodium 45 Rh 102.91
6	Cesium 55 Cs 132.91	Barium 56 Ba 137.33	Lanthanum 57 La 138.91	Hafnium 72 Hf 178.49	Tantalum 73 Ta 180.95	Tungsten 74 W 183.84	Rhenium 75 Re 186.21	Osmium 76 Os 190.23	Iridium 77 Ir 192.22
7	Francium 87 Fr (223)	Radium 88 Ra (226)	Actinium 89 Ac (227)	Rutherfordium 104 Rf (267)	Dubnium 105 Db (268)	Seaborgium 106 Sg (271)	Bohrium 107 Bh (272)	Hassium 108 Hs (270)	Meitnerium 109 Mt (276)

The number in parentheses is the mass number of the longest lived isotope for that element.

Lanthanide series	Cerium 58 Ce 140.12	Praseodymium 59 Pr 140.91	Neodymium 60 Nd 144.24	Promethium 61 Pm (145)	Samarium 62 Sm 150.36	Europium 63 Eu 151.96
Actinide series	Thorium 90 Th 232.04	Protactinium 91 Pa 231.04	Uranium 92 U 238.03	Neptunium 93 Np (237)	Plutonium 94 Pu (244)	Americium 95 Am (243)